Lecture Notes in Mathematics

A collection of informal reports and seminars
Edited by A. Dold, Heidelberg and B. Eckmann, Zürich

263

T. Parthasarathy

Indian Statistical Institute, Calcutta/India

T0222666

Selection Theorems and their Applications

Springer-Verlag
Berlin · Heidelberg · New York 1972

AMS Subject Classifications (1970): 49 A 35, 54 C 65, 90 D 15

ISBN 3-540-05818-4 Springer-Verlag Berlin · Heidelberg · New York
ISBN 0-387-05818-4 Springer-Verlag New York · Heidelberg · Berlin

This work is subject to copyright. All rights are reserved, whether the whole or part of the material is concerned, specifically those of translation, reprinting, re-use of illustrations, broadcasting, reproduction by photocopying machine or similar means, and storage in data banks.

Under § 54 of the German Copyright Law where copies are made for other than private use, a fee is payable to the publisher, the amount of the fee to be determined by agreement with the publisher.

© by Springer-Verlag Berlin · Heidelberg 1972. Library of Congress Catalog Card Number 72-78192. Printed in Germany.

Offsetdruck: Julius Beltz, Hemsbach/Bergstr.

PREFACE

This volume of lecture notes contains results on selection theorems and their applications. Some of the material of this volume had been given as seminar talks at Case Western Reserve University during 1969 - 1970. This volume contains nine sections each of them followed by selected references. We hope this volume will be profitable to specialists in game theory, dynamic programming, control theory, mathematical economics as well as to all mathematicians interested in this area of Mathematics.

I wish to express my sincere thanks to the following Professors: Henry Hermes, Marc Jacobs, Ashok Maitra and Sam Nadler Jr., for several useful suggestions. A particular measure of gratitude is due to Mr. Arun Das who patiently and accurately prepared the final typescript of this volume.

My wife Ranjani proof-read the manuscript. To her I owe my heart-felt thanks.

Finally I wish to express my gratitude to the Indian Statistical Institute for providing the excellent research facilities and to Springer - Verlag for undertaking the publication of these notes.

T. Parthasarathy
Indian Statistical Institute

November 20, 1971

CONTENTS

This volume is dedicated to my parents.

INTRODUCTION

The purpose of these notes is to prove a few selection theorems and to mention some applications of these theorems. We will now briefly outline the contents of these notes. These notes are divided into nine sections.

We start with a selection theorem due to Michael, in section one, which yields a characterisation of paracompactness. In section 2 we are concerned with the following question. Suppose there exists a continuous selection on 2^X (= space of all nonempty closed subsets of some topological space X), then what can you say about X? In other words for what space X does there exist a continuous section on 2^X?

The aim of section three is to show how the existence of continuous selection for certain set-valued maps leads to the existence of classical solution of some generalised differential equations. In section four, we prove a selection theorem due to Dubins and Savage and then we apply this theorem to prove the existence of optimal stationary strategies for the two players in zero-sum two-person stochastic games.

In section five we establish the following result due to Kuratowski and Ryll-Nardzewski. Let Y be a complete separable metric space. Let $F : X \to 2^Y$ be a measurable map [That is, $\{x : F(x) \cap G \neq \emptyset\} \in S$ whenever G is open in Y and S is the countably additive family induced by a field of subsets of X]. Then there is a selector $f : X \to Y$ such that $f^{-1}(G) \in S$ whenever G is open in Y. We also deduce a few more selection theorems with the help of this theorem. In section six, assuming the continuum hypothesis, we present an example (due to Orkin) of a non-analytic subset of [0,1] which is a Blackwell space. This example depends on a measurable selection theorem proved in the previous section.

In section seven we prove essentially the following measurable choice theorem due to von-Neumann. Let $T = [0,1]$ and X be an arbitrary complete separable metric space. If F is an analytic set-valued function from T to X, then there is a Lebesgue measurable point-valued function $f : T \to X$ such that $f(t) \in F(t)$ for almost all t. We use this result while characterising extreme points of sets of vector functions.

Given a set E in the cartesian product $X \times Y$ of two spaces X and Y, a set U is said to uniformise E, if the projections $\pi_X E$ and $\pi_X U$ of E and U through Y onto X coincide, and if, for each $x \in \pi_X E$ the set $(\{x\} \times Y) \cap U$ of points of Y lying above x consists of a single point. In section eight, we are concerned with the question of the existence of such uniformising sets.

In the last section we mention further results on selection theory with some remarks.

1. CONTINUOUS SELECTIONS

One of the most interesting and important problems in topology is the extention problem. Two topolgical spaces X and Y are given, together with a closed subset A of X, and we would like to know whether every continuous function $g : A \to Y$ can be extended to a continuous function f from X (or at least from some open $U \supset A$) into Y. Sometimes there are additional requirements on f, which frequently take the following form : for every $x \in X$, $f(x)$ must be an element of a preassigned subset of Y. This new problem, which we call the selection problem, is clearly more general than the extention problem, and presents a challenge even when A is the null set or a one-point set (where the extention problem is trivial).

Let X and Y denote topological spaces and 2^Y denote the family of non-empty subsets of Y. If $\Phi : X \to 2^Y$ then a selection for Φ is a continuous $f : X \to Y$ such that $f(x) \in \Phi(x)$ for every $x \in X$.

<u>Example 1.1</u>: Let $u : Y \to X$ be onto. Define $\Phi : X \to 2^Y$ by $\Phi(x) = u^{-1}(x)$. Then f is a selection for Φ if and only if f is continuous and $f(x) \in u^{-1}(x)$ for every $x \in X$.

<u>Example 1.2</u>: Let $\psi: X \to 2^Y$, let $A \subset X$ and g be a selection for $\psi|A$. Define $\Phi : X \to 2^Y$ by $\Phi(x) = \{g(x)\}$ if $x \in A$ and $\Phi(x) = \psi(x)$ if $x \in X-A$. Then $f : X \to Y$ is a selection for Φ if and only if f is a selection for ψ which extends g.

The selection problem of the first paragraph can now be rephrased as follows : Under what conditions on X, $A \subset X$, Y and $\Phi : X \to 2^Y$ can every selection for $\Phi|A$ be extended to a selection for Φ, or at least for $\Phi|U$ for some open set $U \supset A$?

The purpose of this section is to give a solution to the above problem.

Definition: Call $\Phi : X \to 2^Y$ lower semicontinuous (abbreviated as l.s.c.) if $\{x : \Phi(x) \cap U \neq \emptyset\}$ is open in X for every open set $U \subset Y$.

The first step in answering the question, is provided by the following elementary but important necessary condition.

Proposition 1.1: If $\Phi : X \to 2^Y$ has the property that, for every $x_0 \in X$, there exists a selection for $\Phi | U$ (U a neighbourhood of x_0) which has a preassigned value $y_0 \in \Phi(x_0)$ at x_0, then Φ is lower semicontinuous.

Example 1.1*: Φ is l.s.c. if and only if u is open.

Example 1.2*: If ψ is l.s.c., A is closed in X and g is continuous, then Φ is l.s.c.

Proof of proposition 1.1: Let V be open in Y; we must show that $G = \{x : \Phi(x) \cap V \neq \emptyset\}$ is open in X. For each $x_0 \in G$, pick a $y_0 \in \Phi(x_0)$; then by assumption, there exists a selection f_0 for $\Phi | U_0$, for some neighbourhood U_0 of x_0 such that $f(x_0) = y_0$. Now if $U_0^* = U_0 \cap \{x : f_0(x) \in V\}$, then U_0^* is a neighbourhood of x_0 which is contained in G. Hence G is open and the proof is complete.

One can easily prove the following facts :

(i) If $\Phi : X \to 2^Y$ is l.s.c. and if $\quad : X \to 2^Y$ is such that $\overline{\psi(x)} = \overline{\Phi(x)}$ for every x where $\overline{\psi(x)}$ and $\overline{\Phi(x)}$ denote the closure of $\psi(x)$ and $\Phi(x)$ respectively, then ψ is l.s.c.

(ii) If $\Phi : X \to 2^Y$ is l.s.c. and if U is an open subset of Y with the property that $\Phi(x) \cap U$ is non-empty for every $x \in X$, then the function $\theta : X \to 2^Y$ defined by $\theta(x) = U \cap \Phi(x)$ is l.s.c.

(iii) If Y is a topological linear space, and if $\Phi : X \to 2^Y$ is l.s.c., then the function $\psi : X \to 2^Y$, defined by $\psi(x)$=convex-hull

Now we are set to prove the following important result due to Michael which gives a characterisation of paracompact spaces. Recall X is paracompact if it is Hausdorff and if every open covering of X has an open locally finite refinement.

Theorem 1.1: The following properties of a T_1-space X are equivalent [1].

(a) X is paracompact.

(b) If Y is a Banach space then every l.s.c. function $\phi: X \to \mathcal{F}_1(Y)$ (family of all closed convex subsets of Y) admits a selection.

First we shall establish the following lemma.

Lemma 1.1: If X is paracompact, Y a normed linear space, ψ a lower semi-continuous function from X to the non-empty convex subsets of Y and $r > 0$, then there exists a continuous $f : X \to Y$ such that $f(x) \in S_r(\psi(x))$ for every $x \in X$, where $S_r(\psi(x)) = \{y: d(y, \psi(x)) < r\}$.

Proof: Let $U_y = \{x : y \in S_r(\psi(x))\}$ for $y \in Y$. Then

$$U_y = \{x : \psi(x) \cap S_r(y) \neq \emptyset\}.$$

Since ψ is l.s.c., U_y is open for every y. [Here $S_r(y)$ stands for the open sphere of radius r with centre y]. Clearly $\{U_y : y \in Y\}$ is a covering for X and hence, since X is paracompact, it has an open locally finite refinement $\{V_\alpha : \alpha \in A\}$. [We assume $V_\alpha \neq V_\beta$ if $\alpha \neq \beta$]. By a theorem of Dieudonne, there exists an open covering $\{W_\alpha : \alpha \in A\}$ such that $\overline{W}_\alpha \subset V_\alpha$. Now we shall construct a family $\{p_\alpha : \alpha \in A\}$ of continuous functions from X to the closed unit interval with the following properties : (i) p_α vanishes outside V_α and (ii) $\sum_{\alpha \in A} p_\alpha(x) = 1$ for every $x \in X$. By the normality of X we can find continuous functions $q_\alpha : X \to [0,1]$ such that $q_\alpha(x)=1$ when $x \in \overline{W}_\alpha$ and $q_\alpha(x) = 0$ when $x \in X-V_\alpha$. Define

$$p_\alpha(x) = \frac{q_\alpha(x)}{\sum_{\alpha \in A} q_\alpha(x)} .$$

Then $\{p_\alpha\}$ satisfies all our requirements. Pick for each $\alpha \in A$, a $y(\alpha) \in Y$ such that $V_\alpha \subset U_{y(\alpha)}$. Define

$$f(x) = \sum_{\alpha \in A} p_\alpha(x)\, y(\alpha).$$

To see that this works, observe first that each $x \in X$ has a neighbourhood U intersecting only finitely many V_α and on this U, f is the sum of finitely many continuous functions. Hence every $x \in X$ has a neighbourhood on which f is continuous and therefore f is continuous on X. Moreover, for each $x \in X$, $f(x)$ is a convex combination of finitely many $y(\alpha)$, all of which lie in the convex set $S_r(\psi(x))$ and hence $f(x) \in S_r(\psi(x))$. This completes the proof of the lemma.

<u>Proof of theorem 1.1</u> : We will prove first (a) $\Rightarrow$ (b). To prove this it is sufficient if we construct a sequence of continuous functions $f_n : X \to Y$ such that for every $x \in X$,
(i) $f_n(x) \in S_{r_{n-2}}(f_{n-1}(x))$ for $n = 2,3,\ldots$ and (ii) $f_n(x) \in S_{r_n}(\Phi(x))$ for $n = 1,2,\ldots,$ where $r_n = 1/2^n$. This will be sufficient, because then the $\{f_n\}$ is uniformly Cauchy and hence converges uniformly to a continuous f and that it follows $f(x) \in \Phi(x)$ for every $x \in X$.

We construct $\{f_n\}$ by induction. The existence of a function f_1 satisfying (ii) follows from lemma 1.1. Suppose $f_1, f_2, \ldots, f_n$ have been constructed to satisfy (i) and (ii) and let us construct f_{n+1} to satisfy (i) and (ii).

For each $x \in X$, let $\Phi_{n+1}(x) = \Phi(x) \cap S_{r_n}(f_n(x))$. Then $\Phi_{n+1}(x)$ is never empty by induction hypothesis. Let us check that Φ_{n+1} is l.s.c. That is, we must show $V=\{x : \Phi_{n+1}(x) \cap U \neq \emptyset\}$ is open if U is open. To do this, we will show that every $x_o \in V$ has a neighbourhood contained in V. Now for $x_o \in V$, pick $y_o \in \Phi(x_o)$ and a positive $\lambda < r_n$ such that $y_o \in S_\lambda(f_n(x_o))$. Let

$$W_1 = \{ x : \Phi(x) \cap S_\lambda(f_n(x_o)) \cap U \neq \emptyset \},$$

$$W_2 = \{ x : f_n(x) \in S_{r_n-\lambda}(f_n(x_o))\} .$$

Clearly W_1 is open since Φ is l.s.c. and W_2 is open for f_n is continuous; and $x_o \in W_1 \cap W_2 \subset V$. Since Φ_{n+1} is l.s.c. We can apply lemma 1.1 to find a continuous $f_{n+1} : X \to Y$ such that $f_{n+1}(x) \in S_{r_{n+1}}(\Phi_{n+1}(x))$ for every x. But then $f_{n+1}(x) \in S_{r_{n-1}}(f_n(x))$ and $f_{n+1}(x) \in S_{r_{n+1}}(\Phi(x))$. This completes the proof of (a) $\Rightarrow$ (b).

Now we shall prove (b) $\Rightarrow$ (a). To show this, it is enough if we establish that every open covering $\mathcal{U}$ of X has a partition of unity P subordinated to it; this means that there exists a collection P of continuous functions from X to the non-negative reals such that $\sum_{p \in P} p(x) = 1$ for every $x \in X$ and every $p \in P$ vanishes outside some $U \in \mathcal{U}$. Let $\mathcal{U}$ be an open covering of X. Let $Y = \ell_1(\mathcal{U}) = \{ y \mid y : \mathcal{U} \to R, \sum_{U \in \mathcal{U}} |y(U)| < \infty \}$. Here $\|y\| = \sum_{U \in \mathcal{U}} |y(U)|$. Clearly Y is a Banach space. Let $C = \{ y : y \in Y, y(U) \geq 0$ for every $U \in \mathcal{U}$ and $\sum y(U) = 1 \}$. Plainly C is a closed convex subset of Y. Now for $x \in X$, let $\Phi(x) = C \cap \{ y : y \in Y, y(U) = 0$ for every $U \in \mathcal{U}$ with $x \notin U \}$. Clearly $\Phi(x) \in \mathcal{F}(Y)$ for every $x \in X$. We will now show that Φ is l.s.c.

Let us first of all show that, for every $y \in C$ and $\varepsilon > 0$, there exists a $y' \in C$ such that $\|y-y'\| < \varepsilon$ and $y'(U) > 0$ for only finitely many $U \in \mathcal{U}$. To find such a y', we need only pick $U_1, U_2, \dots U_n \in \mathcal{U}$ such that $y(U_i) > 0$ for all i and $y(U_1) + y(U_2) + \dots + y(U_n) = \delta > 1 - \varepsilon/2$, and then define $y' \in C$ by $y'(U) = 0$ if $U \notin \{ U_1, U_2, \dots U_n \}$, $y'(U_1) = y(U_1) + 1-\delta$ and $y(U_i)=y'(U_i)$ for $i = 2,3,\dots n$. It is easy to see that $\|y-y'\| \leq 2(1-\delta) < \varepsilon$, and therefore y' satisfies all our requirements. Now we will show that Φ is l.s.c. To show this it is enough if we establish the following : if $x \in X$, $y \in \Phi(x)$, and $\varepsilon > 0$, then there exists a neighbourhood U of x in X such that, for every $x' \in U$, there exists a $y' \in \Phi(x')$ with $\|y-y'\|<\varepsilon$. Suppose, therefore, that $y \in \Phi(x)$ and $\varepsilon > 0$ are given, and let y' and $U_1, U_2, \dots U_n$ be as given above. Let $U = U_1 \cap \dots \cap U_n$. Since $y(U_i) > 0$ for $i = 1,2,\dots n$, it follows from the definition of Φ that $x \in U_i$ for $i = 1,2,\dots n$ and hence U is indeed a neighbourhood of x. It also follows from the definition of Φ that $y' \in \Phi(x')$ for every $x' \in U$, and therefore U satisfies all our requirements.

By assumption (b), there now exists a selection f for Φ. For each $U \in \mathcal{U}$, define $f_U : X \to R$ by $f_U(x) = [f(x)]\,(U)$. It now follows immediately from the definitions that $\{ f_U : U \in \mathcal{U}\}$ is a partition of unity on X, and this partition is subordinated to $\mathcal{U}$, since f_U vanishes outside U for every $U \in \mathcal{U}$. This completes the proof of the theorem.

Remark 1.1 : Even if X is the closed unit interval in the above theorem, (b) becomes false if Banach is replaced by 'normed linear' as the following example shows.

<u>Example 1.3</u> : Let X be the closed unit interval. Then there exists a separable normed linear space and a l.s.c. function $\Phi : X \to \mathcal{F}(Y)$ (= closed convex subsets of Y) for which there is no selection.

<u>Proof</u>: Let Z be the set of rationals in X, and suppose Z is ordered as a sequence $z_1, z_2, \ldots$. Let $Y = \{y : y \in \ell_1(Z),$ $y(x) \neq 0$ for only finitely many $x \in Z\}$. Let $C = \{y : y \in Y,$ $y(x) \geq 0$ for all $x \in Z\}$. Finally let

$$\Phi(x) = \begin{cases} C & \text{if } x \in X - Z \\ C \cap \{ y : y \in Y, \ y(z_n) \geq 1/n \} & \text{if } x = z_n. \end{cases}$$

It is easy to check that Φ is l.s.c. Let us show that there is no selection for Φ.

Suppose f were a selection for Φ ; since f is continuous, each $z_n \in Z$ has a neighbourhood U_n in X such that $[f(x)](z_n) > 1/2n$ whenever $x \in \overline{U}_n$. By induction pick a sequence $\{n_k\}$ of distinct integers such that $z_{n_{k+1}} \in \bigcap_{i=1}^{k} U_{n_i}$ for all k. Then $\{\overline{U}_{n_k}\}$ is a sequence of closed subsets of X with finite intersection property, and hence there is an $x_0 \in \bigcap_{k=1}^{\infty} U_{n_k}$. But then $[f(x_0)](z_{n_k}) > 0$ for all k, which is impossible.

As applications of the above theorem we cite the following two propositions :

<u>Proposition 1.2</u> : Let X be a paracompact space, Y a Banach space and $\Phi : X \to \mathcal{F}(Y)$, a l.s.c. function. Let $m(x) = \inf \{ \|y\| : y \in \Phi(x) \}$ and suppose that $p : X \to R$ is l.s.c. with $p(x) \geq 0$ for all x, and $p(x) > m(x)$ whenever $m(x) > 0$. Then there exists a selection f for Φ such that

$\| f(x) \| \leq p(x)$ for every $x \in X$.

Proof: Let $\psi(x) = \begin{cases} \Phi(x) \cap \{ y : y \in Y, \| y \| < p(x)\} & \text{if } p(x) > 0 \\ \{0\} & \text{if } p(x) = 0. \end{cases}$

It is not hard to check that ψ is l.s.c. Hence if we define $\Theta : X \to \mathcal{F}(Y)$ by $\Theta(x) = \overline{\psi(x)}$ then Θ is also l.s.c. Hence by the above theorem, there now exists a selection f for Θ and f satisfies all our requirements.

Proposition 1.3 : A topological space X is paracompact if and only if it is dominated by a collection of paracompact subsets.

Before proving proposition 1.3, we will start with a definition.

Definition : Let X be a topological space and $\mathcal{B}$ a collection of closed subsets of X. Then $\mathcal{B}$ dominates X if, whenever $A \subset X$ has a closed intersection with every element of some subcollection $\mathcal{B}_1$ of $\mathcal{B}$ which covers A, then A is closed.

Remark 1.2 : If $\mathcal{B}$ dominates X and if $\mathcal{B}_1 \subset \mathcal{B}$ then $\cup \mathcal{B}_1$ is closed. If Y is any topological space and if $f : \cup \mathcal{B}_1 \to Y$ is a function such that $f|B$ is continuous for every $B \in \mathcal{B}_1$, then f is continuous.

Proof of proposition 1.3 : The ''only if'' part is obvious, since we need only take $\mathcal{B} = \{ X \}$. Let us therefore prove the ''if'' part. By theorem 1.1, it is sufficient to prove that if Y is a Banach space, and $\Phi : X \to \mathcal{F}(Y)$ is l.s.c., then Φ admits a selection. So let Y and Φ be given, and let us find f.

Consider the class $\mathcal{A}$ of all couples of the form $(\mathcal{C}, h)$ where $\mathcal{C} \subset \mathcal{B}$, and h is a selection for $\Phi \mid \cup$. We partially order $\mathcal{A}$ in the obvious manner. From the remark made above, it follows that every simply ordered sub-class of $\mathcal{A}$ has an obvious upperbound, and hence, by Zorn's lemma, $\mathcal{A}$ has a maximal element say $(\mathcal{C}_o, h_o)$. We need only show that $\mathcal{C}_o = \mathcal{B}$, for then we can simply take $f = h_o$. Suppose $\mathcal{C}_o \neq \mathcal{B}$; then there exists a $B \in \mathcal{B}$ with $B \notin \mathcal{C}_o$. Let $B \cap (\cup \mathcal{C}_o) = B^*$. Now $h_o | B^*$ is a selection for $\Phi \mid B^*$ and hence by theorem 1.1 and lemma 1.2 (stated and proved below) it can be extended to a selection g for $\Phi \mid B$. If we now let $\mathcal{C}_1 = \{B\} \cup \mathcal{C}_o$ and define $h_1 : \cup \mathcal{C}_1 \to Y$ by $h_1(x) = h_o(x)$ if $x \in \cup \mathcal{C}_o$ and $h_1(x) = g(x)$ if $x \in B$, then $(\mathcal{C}_1, h_1)$ is an element of $\mathcal{A}$ which is larger than $(\mathcal{C}_o, h_o)$. This contradicts the maximality of $(\mathcal{C}_o, h_o)$ and thus the proof is complete.

<u>Lemma 1.2</u> : If $\mathcal{S} \subset 2^Y$ contains all one-point subsets of elements of $\mathcal{S}$, then the following two properties of $\mathcal{S}$ are equivalent.

(a) Every lower semicontinuous $\Phi : X \to \mathcal{S}$ admits a selection.

(b) If $\Phi : X \to \mathcal{S}$ is l.s.c., then every selection for $\Phi \mid A$ (where A is a closed set) can be extended to a selection for Φ.

<u>Proof</u> : That (b) $\Rightarrow$ (a) is obvious. To show that (a) $\Rightarrow$ (b), let $\Psi : X \to \mathcal{S}$ be l.s.c., let $A \subset X$ be closed and let g be a selection for Ψ. Let $\Phi : X \to \mathcal{S}$ be defined as follows.

$$\Phi(x) = \{g(x)\} \quad \text{for } x \in A$$

$$= \Psi(x) \quad \text{for } x \in X-A.$$

It is not hard to check that Φ is l.s.c., hence admits a selection by assumption (a) and this selection has the required properties and the proof is thus complete.

For the proper understanding theorem 1.1 and theorem 1.2 (stated below without proof) the reader should keep in mind the following consequence of lemma 1.2, where Y is called an extention space with respect to X if, for every closed $A \subset X$, every continuous $g : A \to Y$ can be extended to a continuous $f : X \to Y$. If $\mathcal{Y} \subset 2^Y$ contains every one point subset of Y and also Y itself, then (a) $\Rightarrow$ (b) below :

(a) Every l.s.c. $\Phi : X \to \mathcal{Y}$ admits a selection.

(b) Y is an extention space with respect to X.

Theorem 1.2 : If X is paracompact and zero-dimensional and if Y is a complete metric space, then every lower semicontinuous function Φ from X to the non-empty, closed subsets of Y admits a selection.

Proof of theorem 1.2 is similar in nature to the proof of theorem 1.1 and hence omitted. [Recall that X is zero-dimensional if every finite open covering of X has a disjoint finite open refinement, in other words, X has a base consisting of sets which are both open and closed].

The significance of theorem 1.1 lies in the fact that it is the first characterisation of paracompactness which deals with continuous functions rather than coverings. Further the theorem yields the fact that every paracompact space and hence every metric space has property (b) of theorem 1.1.

REFERENCES

[1] E. Michael, Continuous selections I, Ann. of Math. 63 [1956], 361-382.

[2] E. Michael, A survey of continuous selections, Springer-Verlag Lecture notes in Mathematics, Edited by W. M. Fleischman, on ''Set valued mappings, selections and topological properties of 2^X'' . 171 [1970], 54-58.

2. CONTINUOUS SELECTIONS ON METRIC CONTINUUM

In this section we are concerned with the following problem : X is any topological space and 2^X is the space of all nonempty closed subsets of X with the Vietoris topology. Let S be a subspace of 2^X. A function $f : S \to X$ is called a selection on S if $f(E) \in E$ for every $E \in S$. Suppose there exists a continuous selection on 2^X, then what can you say about X ? In other words for what spaces X does there exist a continuous selection on 2^X ?

Throughout this section we assume X to be a metric continuum unless otherwise stated. By 'continuum' we mean compact, connected, Hausdorff. $F_2(X)$ will stand for all two-element and one element sets of X. A subbase for the Vietoris topology on X consists of all sets having one of the following forms :

$$\{F : F \in 2^X, \ F \cap G \neq \emptyset\} \text{ and } \{F : F \in 2^X, \ F \subset G\}$$

where G is an arbitrary open set in X. But if X is a compact metric space, the Hausdorff metric on 2^X induces the Vietoris topology. The Hausdorff metric d on 2^X is defined as follows

$$d(A,B) = \max [\sup_A \rho(x,B), \ \sup_B \rho(x,A)]$$

where $A,B \in 2^X$ and ρ is the metric on X. Also, X is compact if and only if 2^X is compact [1]. The purpose of this section is to prove the following theorem [2].

<u>Theorem 2.1</u>: If M is a metric continuum for which there is a continuous selection on $F_2(M)$ then M is an arc.

<u>Remark 2.1</u>: Note that if X is a continuum and if there is a continuous selection on $F_2(X)$, then there is also a continuous

selection on 2^X. However this does not remain true in the more general case, for example in the space of reals [3].

Definition : If S is a set and suppose f is a selection on $F_2(S)$. An element $a \in S$ is an end element of S with respect to f provided that either each element $x \in S - \{a\}$, $f(\{a,x\}) = a$ or for each $x \in S - \{a\}$, $f(\{a,x\}) = x$.

Lemma 2.1: If S is a set and f is a selection on $F_2(S)$, then S has at most two end-elements with respect to f.

Proof : Suppose a_1, a_2, a_3 are three end-elements. Then for $i = 1,2,3$ either

$$(i,1) \ldots \quad f(a_i,\ x) = a_i$$

or

$$(i,2) \ldots \quad f(a_i,\ x) = x\ .$$

Suppose that $(1,1)$ holds. Then for $j = 2,3$, $f(a_1, a_j) = a_1$, so that $(j,1)$ cannot hold and $(j,2)$ must. Consider $f(a_2, a_3)$. Since $(2,2)$ holds this cannot be a_2 and since $(3,2)$ holds this cannot be a_3. But it must be one or the other. Suppose on the other hand that $(1,2)$holds.Then by a similar argument both $(2,1)$ and $(3,1)$ hold. But then $f(\{a_2, a_3\})$ is both a_2 and a_3, which is also impossible.

Lemma 2.2 : Let S be T_2 and connected, and suppose f is a continuous selection on $F_2(S)$. Then each non-end element c of S is a cut-point of S (that is, $S - \{c\}$ is not connected).

Proof: Let H be the set of all points x in $S - \{c\}$ such that $f(\{x,c\}) = x$ and K be the set of all points y in $S - \{c\}$ such that $f(\{y,c\}) = c$. Then neither H nor K is empty, since otherwise c would be an end element ; further H and K are disjoint. We shall now show, in order to complete the proof of the lemma, that c is a cut-point of S. To do this it is enough if we show H and K are open. Let $x \in H$. By continuity of f, there is an open set U containing x such that if $y \in U$, $f(\{y,c\}) = y$. Hence $U \subset H$ so that H is open. Similarly K is open.

Proposition 2.1 : Let S be T_2 (= Hausdorff) and connected and suppose f is a continuous selection on $F_2(S)$. Then S contains at most two non-cut points.

Proof : This follows from lemma 2.1 and 2.2.

Proposition 2.2 : Every continuum K of more than one point has at least two non-cut points.

Proposition 2.3 : If K is a metric continuum with exactly two non-cut points then K is homeomorphic to the unit interval I, that is, K is an arc.

Proof of theorem 2.1 : The proof is immediate from propositions 2.1, 2.2 and 2.3. Of course, we have yet to establish propositions 2.2 and 2.3. The rest of this section is devoted to the proof of them. We will start with some preliminaries.

Definition : A simple chain connecting two points a and b of a space X is a sequence $U_1, U_2, \ldots U_n$ of open sets of X such that $a \in U_1$ only, $b \in U_n$ only and $U_i \cap U_j \neq \emptyset$ whenever $|i-j| \leq 1$.

Definition : A space X is pathwise connected if for any two points x and y in X, there is a continuous function $f : [0,1] \to X$ such that $f(0) = x$ and $f(1) = y$. The function f (as well as its range) is called a path from x to y.

Definition : We call X arcwise connected, for any two points x and y in X if there is a homeomorphism $f : [0,1] \to X$ such that $f(0) = x$ and $f(1) = y$. The function f (as well as its range) is called an arc from x to y.

At this stage we would like to make the following observations: For a proof of these refer [4].

(i) If X is connected and $\mathcal{U}$ is any open cover of X, then any two points of a and b of X can be connected by a simple chain consisting of elements of $\mathcal{U}$.

(ii) Every pathwise connected space is connected.

(iii) Every connected, locally pathwise connected (that is each point has a neighbourhood base consisting of pathwise connected sets) X is pathwise connected.

<u>Definition</u>: A continuum K (compact, connected T_2-space) in X is irreducible about a subset A of X provided $A \subset K$ and no proper subcontinuum of K contains A. If $A = \{a, b\}$, we say K is irreducible between a and b.

One can show that if K is any continuum, any subset A of K lies in a subcontinuum irreducible about A using Zorn's lemma.

<u>Definition</u>: Let X be a connected T_1-space. A cut point of X is a point $p \in X$ such that $X - \{p\}$ is not connected. If p is not a cut point of X, we call it a non-cut point of X. A cutting of X is a set $\{p, U, V\}$ where p is a cut point of X and U and V disconnect $X - \{p\}$ [That is, U and V are disjoint non-empty open subsets of X whose union is $X - \{p\}$.

The property of being a cut point (in fact, of being a cutting) is preserved under homeomorphism, but continuous maps can destroy cut points. For example consider the map $f(x) = (\cos x, \sin x)$ of $[0, 2\pi]$ onto the unit circle in R^2.

<u>Lemma 2.3</u>: If K is a continuum and $\{p,U,V\}$ is a cutting of K then $U \cup \{p\}$ and $V \cup \{p\}$ are connected (and thus are continuum).

<u>Proof</u>: It suffices to prove the lemma for $U \cup \{p\}$. But the map f defined on K by

$$f(x) = \begin{cases} x & \text{if } x \in U \cup \{p\} \\ p & \text{if } x \in V \end{cases}$$

carries K onto $U \cup \{p\}$, and f is continuous on each of the closed sets $U \cup \{p\}$ and $V \cup \{p\}$, so f is continuous. Thus $U \cup \{p\}$ is the continuous image of a connected space and therefore connected. [Since $U \cup \{p\} = K-V$, $U \cup \{p\}$ is closed in K and thus compact. The part of the theorem in parentheses follows].

<u>Lemma 2.4</u>: If K is a continuum and $\{p,U,V\}$ is a cutting of K, then each of U and V contains a non-cut point of K.

<u>Proof</u>: Suppose each point x in U is a cut point inducing a cutting $\{x, U_x, V_x\}$ of K. If both U_x and V_x meet $V \cup \{p\}$, they disconnect $V \cup \{p\}$ which is impossible by lemma 2.3. So one, say U_x is contained in U. Now $U_x \cup \{x\}$ is a continuum for each $x \in U$ by lemma 2.3. Since $\{U_x \cup \{x\} : x \in U\}$ is directed by inclusion, $\bigcap_{x \in U} [U_x \cup \{x\}]$ is a nonempty continuum contained in U.

Pick $q \in \bigcap_{x \in U} [U_x \cup \{x\}]$. Then $U_q \subset U$ (as above) and if $r \in U_q$, then $q \notin U_r$ [otherwise U_r and V_r both meet $V_q \cup \{q\}$ and disconnect it]. Then $U_r \cup \{r\}$ does not contain q which is a contradiction.

<u>Proof of proposition 2.2</u>: If p is a cut point of K, then a cutting (p,U,V) of K exists, and each of U and V contains a

non-cut point of K, by the previous lemma. On the other hand, if no cut point of K exists, certainly there are two non-cut points and the proof is complete.

Let N be the set of non-cut points of a continuum K and suppose a proper subcontinuum L of K contains N. [We will show that this is impossible]. If $x \in K-L$, then a cutting $\{x, U, V\}$ of K exists and L must lie in one or the other of U and V say $L \subset U$. Then $V \cup \{x\}$ being a continuum itself has two noncut points and thus a noncut point $y \neq x$. Then $[V \cup \{x\}] - \{y\}$ is connected, and $U \cup \{x\}$ is connected and these sets meet, so their union is connected. But their union is $K - \{y\}$, while $y \in V$, hence $y \notin U$, hence not in L ; this is a contradiction since L contains all the noncut points of K. Thus we have shown that a continuum K is irreducible about the set of its non-cut points.

<u>Definition</u>: A cut point p in a connected space X separates a from b if a cutting $\{p, U, V\}$ exists with $a \in U$ and $b \in V$. The set consisting of a,b and all points p which separate a and b is denoted by E(a,b). The separation order on E(a,b) is defined by : $p_1 \leq p_2$ if $p_1 = p_2$ or p_1 separates a from p_2. This is easily seen to be a partial order on E(a,b). In fact, the separation order on E(a,b) is a linear order ! (Prove this).

<u>Lemma 2.5</u>: (a) If E(a,b) has more than two points, its order topology is weaker than its subspace topology.

(b) If K is a continuum with exactly two non-cut points a and b, then E(a,b) = K and the topology on K is the order topology.

Proof: (a) It suffices to note that, for $p \in E(a,b)$, the sets $U_p \cap E(a,b)$ and $V_p \cap E(a,b)$ are open in $E(a,b)$ and

$$U_p \cap E(a,b) = \{q : q \in E(a,b),\ q < p\}$$

$$V_p \cap E(a,b) = \{q : q \in E(a,b),\ p < q\}$$

(b) If $p \in K$ and p is not one of a or b, then given any cutting $\{p, U, V\}$ of K, by lemma 2.4 U and V each contain one of a and b. Thus $p \in E(a,b)$, so $E(a,b) = K$. From (a), the order topology is weaker than the given topology on K. Suppose conversely, that U is open in K and $p \in U$. First assuming that p is not one of a or b, we will show that U contains some interval $(r,s) = \{q : r < q < s\}$ containing p. If not, then whenever $p \in (r,s)$, the closed interval $[r, s] = \{q : r \leq q \leq s\}$ meets $K-U$. But the sets $[r,s] \cap K-U$ then form a family of closed subsets of K with the finite intersection property (each $[r, s]$ is closed in K by part (a)). Thus their intersection (in the compact space K) is nonempty. But $p \in U$ and

$$\cap \{[r, s] : p \in [r, s]\} = \{p\}$$

which leads to a contradiction. If p is one of a or b, the argument is similar.

Remark 2.2 : Every continuum with exactly two non-cut points is a totally ordered set with the order topology induced by its separation order.

Definition: Let X and Y be ordered spaces. A map f of X onto Y is an order isomorphism if f is 1-1 and $x < y \iff f(x) < f(y)$.

Every order isomorphism is homeomorphism relative to the order topologies on X and Y. Let P denote the set of dyadic rationals in $(0,1)$; that is, P consists of all numbers of the form $K/2^n$ for $n = 1,2,\ldots$ and $K = 1,2,\ldots,2^n-1$. Then it is easy to see that P has no largest or smallest element and if $p, q \varepsilon P$ with $p < q$, then for some $r \varepsilon P$, $p < r < q$. Let D be any countable linearly ordered set with the following property (i) It has no largest or smallest element (ii) If $p, q \varepsilon D$ with $p < q$ then there exists a $r \varepsilon D$ with $p < r < q$. Then D is order isomorphic to P.

Proof of proposition 2.3: Let D be a countable dense subset of K not containing the non-cut points a and b. Note that:

(a) D has no smallest or largest element

(b) given $p,q \varepsilon D$ with $p < q$ there is an element $r \varepsilon D$ with $p < r < q$.

From the remarks made above it follows that D is order isomorphic to P, the set of dyadic rationals in $(0,1)$. Let f be an order isomorphism of D onto P. But each point p of K other than a or b is a cut point, dividing K into sets A_p and B_p (i.e. $x < y$ whenever $x \varepsilon A_p$ and $y \varepsilon B_p$). It follows that $f(A_p \cap D)$ and $f(B_p \cap D)$ form a Dedekind cut of the dyadic rationals and thus uniquely determine an element $F(p)$ of $(0,1)$. Defining $F(a) = 0$ and $F(b) = 1$, we have completed the job of extending f to what is obviously an order isomorphism, and thus a homeomorphism of K onto I. This completes the proof of proposition 2.3.

With the aid of theorem 2.1, one can prove the following theorem. For a proof refer [2].

Theorem 2.2 : If M is a locally compact separable metric space for which there is a continuous selection on $F_2(M)$, then M is homeomorphic to a subset of the real line.

We close this section with a final remark, namely, if X is a metrizable continuum, then a continuous selection for 2^X exists if and only if X is an arc.

REFERENCES

[1] K. Kuratowski, Topology, Vol. I, Acad. Press and P.W.N., [1966].

[2] K. Kuratowski, S.B. Nadler, and G.S. Young, Continuous selections on locally compact separable metric spaces, Bulletin De L'Academie Polonaise Des Sciences 18, [1970], 5-11.

[3] E. Michael, Topologies on spaces of subsets, Trans. Amer. Math. Soc. 71, [1951], 152-182.

[4] S. Willard, General Topology, Addison-Wesley Publishing Company, [1970].

3. CONTINUOUS SELECTIONS AND SOLUTIONS OF GENERALISED DIFFERENTIAL EQUATIONS

Let E^n denote Euclidean n dimensional space, B^n a closed, origin centered ball of radius b in E^n, and $\underline{C}(B^n)$ the metric space of all nonempty compact subsets of B^n with the Hausdorff topology. We consider the generalised differential equation

$$\dot{x}(t) \in R(x(t)),\ x(0) = x^o \in E^n \quad \left(\dot{x}(t) = \frac{dx(t)}{dt} \right) \qquad \ldots \qquad (1)$$

where $R : E^n \rightarrow \underline{C}(B^n)$ is continuous. A solution of equation (1) is an absolutely continuous function ϕ such that $\dot{\phi}(t) \in R(\phi(t))$ for almost all t in some neighbourhood of zero, $\phi(0) = x^o$; a classical solution of equation (1) is a continuously differentiable function ϕ with $\phi(0) = x^o$ and $\phi(t) \in R(\dot{\phi}(t))$ for t in some neighbourhood of zero.

The aim of this section is to show how the existence of continuous selection for R leads to (classical) solution of equation (1). For instance, if R has convex values, the existence of classical solutions of equation (1) may be shown in many ways [2]. This may be proved as follows. Let $r(x)$ be the circumcentre for $R(x)$, that is ; the centre of the unique smallest sphere in E^n containing $R(x)$. Then [1, pp. 76-77] r is a continuous selection for R. But $\dot{x}(t) = r(x(t))$, $x(0) = x^o$ has a solution and this solution is a classical solution of (1).

We shall show in this section that if R is of bounded variation, a classical solution of equation (1) exists, while under a weaker condition, termed property A, a solution is shown to exist. These conditions are pertinent to the question of when a continuous map $Q : [0,T] \rightarrow \underline{C}(B^n)$ admits a continuous selection, i.e. a continuous map $q : [0,T] \to E^n$ with $q(t) \varepsilon Q(t)$ for all t. It is shown that if Q is of bounded variation, a continuous selection exists. A special case of this is if Q is Lipschitzian in the Hausdorff metric, say with Lipschitz constant k, then one may show Q admits a Lipschitz continuous selection q, also with Lipschitz constant k.

We assume, throughout, that $R : E^n \to \underline{\underline{C}}(B^n)$ continuously, where B^n is as in the introduction. $C[0,T]$ will denote the space of continuous (n vector valued) functions on $[0,T]$ with the uniform norm, and $C_b[0,T]$ the compact subset of $C[0,T]$ defined by

$$C_b[0,T] = \{ x : x \varepsilon C[0,T],\ x(0) = x^o, |x(t)-x(t')| \leq b|t-t'| \}.$$

If $Q \varepsilon \underline{\underline{C}}(B^n)$ and $y \varepsilon E^n$ we use the notation $\rho(y, Q) = \inf \{ |y-q| : q \varepsilon R \}$, while for $Q_1, Q_2 \varepsilon \underline{\underline{C}}(B^n)$, $h(Q_1, Q_2)$ denotes the Hausdorff distance between these sets.

For $Q : [0,T] \to \underline{\underline{C}}(B^n)$, define the variation of Q on the subinterval $[t - \sigma, t]$, $\sigma > 0$, denoted by $V_{t-\sigma}^t (Q)$, as follows. Let P denote a partition of $[t-\sigma, t]$, i.e. a finite collection of points $t-\sigma = t_o < t_1 < \dots < t_{k+1} = t$, and let $\mathcal{P}$ denote the set of all such partitions. For the partition P, define

$$V_{t-\sigma}^t (Q ; P) = \sum_{\nu=0}^{k} h(Q(t_{\nu+1}), Q(t_\nu)) \quad \text{and}$$

$$V_{t-\sigma}^t (Q) = \sup_{P \varepsilon \mathcal{P}} V_{t-\sigma}^t (Q, P).$$

<u>Property A</u> : For each $x \varepsilon C_b[0,T]$, $V_{t-\sigma}^t (R(x(\cdot))) \varepsilon L_1[0,T]$ and $\lim_{\sigma \to 0} \int_0^T V_{t-\sigma}^t (R(x(.)))\, dt = 0$ uniformly for $x \varepsilon C_b[0, T]$.

(Assume $x(t) = x(t) = x(0)$ for $t < 0$ so that $V_{t-\sigma}^{t}(R(x(\cdot)))$ is defined for $t-\sigma < 0$. This convention will be assumed throughout, when necessary).

At this juncture we would like to make the following observations :

(i) It is clearly possible that the variation $V_{t-\sigma}^{t}(Q)$ is unbounded for some t, yet $V_{t-\sigma}^{t}(Q) \in L_1$. For example, consider the special case Q a point valued function in the interval $[0,1]$ defined by

$$q(t) = \begin{cases} (1-t)\,\mathrm{Sin}(1/1-t) & \text{when } t \neq 1 \\ 0 & \text{when } t = 1. \end{cases}$$

(ii) If R is Lipschitzian, say $h(R(x), R(x')) \leq K|x-x'|$, clearly property A is satisfied. Indeed $V_{t-\sigma}^{t}(R(x(\cdot));P) = \sum_{\gamma=0}^{k} h(R(x(t_{\gamma+1})), R(x(t_{\gamma}))) \leq K\, b \sum_{\gamma=0}^{k}(t_{\gamma+1} - t_{\gamma}) = K\, b\, \sigma$ for $x \in C_b[0,T]$.

(iii) Let $R : E^n \to \underset{\sim}{C}(B^n)$ continuously and let $S = \{y : |y-x^o| \leq b\, T\}$. We define the variation of R in S, denoted $V(R,S)$ as follows. Let λ be any finite collection of points $y^1,\ldots,y^{k+1}$ in S such that $\sum_{\gamma=1}^{k} |y^{\gamma+1} - y^{\gamma}| \leq b\, T$ and let Λ denote the set of all such collections. Let $V(R,S,\lambda) = \sum_{i=1}^{k} h(R(y^{i+1}), R(y^{i}))$ and $V(R,S)$ is defined as $\sup_{\lambda \in \Lambda} V(R,S,\lambda)$. If $V(R,S) < \infty$ we say R has bounded variation in S. Clearly if R has bounded variation in S, then for any $x \in C_b[0,T]$, $V_{t-\sigma}^{t}(R(x(\cdot))) < \infty$; in fact $V_o^t(R(x(\cdot)))$ is finite for all t in $[0,T]$ and continuous as a function of t. In this case, $V_{t-\sigma}^{t}(R(x(\cdot))) = V_o^t(R(x(\cdot))) - V_o^{t-\sigma}(R(x(\cdot)))$ and hence $V_{t-\sigma}^{t}(R(x(\cdot))) \to 0$ as $\sigma \to 0$ uniformly in $x \in C_b[0,T]$.

(iv) For $Q \in \underline{\underline{C}}(B^n)$ let Q^ε denote a closed $\varepsilon > 0$ neighbourhood of Q. For each $x \sigma C_b[0,T]$ we consider $R^\varepsilon(x(.)) : [0,T] \to \underline{\underline{C}}(B^n)$. [increasing the radius of B^n by ε if necessary but retaining the notation B^n]. Then $\{R^\varepsilon(x(.)) : x \in C_b[0,T]\}$ is an equicontinuous family. One may note that in the proof of theorem 3.1, which follows, the essential point is to conclude that one may obtain a continuous (or measurable) selection r_x for each $R^\varepsilon(x(\cdot))$ such that the family $\{r_x : x \in C_b[0,T]\}$ is conditionally compact.

Theorem 3.1 [4]. Let $R : E^n \to \underline{\underline{C}}(B^n)$ continuously. Then,

(a) If R is of bounded variation in the ball $\{y : |y-x^o| \leq b\ T\}$ equation (1) admits a classical solution on $[0,T]$.

(b) If R satisfies property A, equation (1) admits a solution on $[0,T]$.

Proof of (a) : The family $\{R(x(.)) : x \in C_b[0,T]\}$ is equicontinuous where we consider each $R(x(.))$ as a map of $[0,T]$ to $\underline{\underline{C}}(B^n)$. For each $i = 1,2,\ldots$, let $\varepsilon_i > 0$, $\varepsilon_i \to 0$ as $i \to \infty$ and for each ε_i, let $\delta_i > 0$ be such that, $h(R(x(t)), R(x(t'))) < \varepsilon_i$ if $|t-t'| < \delta_i$, $x \in C_b[0,T]$. We assume, without loss of generality, that $\delta_i \to 0$ as $i \to \infty$.

Pick $q_o \in R(x^o)$. For each $i = 1,2,\ldots$ define $r^i : [-\delta_1, T] \to E^n$ as follows. Let $r^i(t) = q_o$ if $-\delta_1 \leq t \leq \delta_i$. Let $r^i(t) = q_o$ if $-\delta_1 \leq t \leq \delta_i$. Define

$$x^i(t) = x^o + \int_0^t r^i(\tau)\, d\tau \qquad \ldots (2)$$

for $t \in [-\delta_1, \delta_i]$, and choose $q_1^i \in R(x^i(\delta_i))$ such that $|q_o - q_1^i| = \rho(q_o, R(x^i(\delta_i)))$. We define $r^i(2\delta_i)$ as q_1^i and extend r^i to $[-\delta_1, 2\delta_i]$ as the straight line segment joining q_o and q_1^i

on the interval $[\delta_1, 2\delta_1]$. We now extend x^1 to $[-\delta_1, 2\delta_1]$ by equation (2). Proceeding inductively, if x^1 is define on the interval $[-\delta_1, j\,\delta_1]$ choose $q_j^1 \in R(x^1(j\delta_1))$ such that $|q_{j-1}^1 - q_j^1| = \rho(q_{j-1}^1, R(x^1(j\delta_1)))$, define $r^1((j+1)\,\delta_1)$ as q_j^1 and extend r^1 to $[-\delta_1, (j+1)\,\delta_1]$ as the straight line segment joining q_{j-1}^1 and q_j^1 on $[j\delta_1, (j+1)\,\delta_1]$. Then extend x^1 to the interval $[-\delta_1, (j+1)\,\delta_1]$ by equation (2). Continue in this manner until each function r^1 and x^1 is defined on $[-\delta_1, T]$. We note that $r^1 \in C[-\delta_1, T]$ while $x^1 \in C_b[0,T]$, when restricted to $[0,T]$. Furthermore we have :

(i) $$|r^1((j+1)\,\delta_1) - r^1(j\delta_1)| = |q_j^1 - q_{j-1}^1| = \rho(q_{j-1}^1, R(x^1(j\delta_1)))$$

$$\leq h(R(x^1((j-1)\,\delta_1)), R(x^1(j\delta_1))) < \varepsilon_1$$

(ii) Given any $t \in [0,T]$ there exists an integer j such that $|j\delta_1 - t| < \delta_1$. Then

$$\rho(r^1(t), R(x^1(t))) \leq |r^1(t) - r^1(j\delta_1)| + \rho(r^1(j\delta_1), R(x^1(j\delta_1)))$$

$$+ h(R(x^1(j\delta_1)), R(x^1(t))) \leq 3\varepsilon_1$$

(iii) For any $t, t' \in [0,T]$, let j, j' be integers such that $|t - j\delta_1| < \delta_1$, $|t' - j'\delta_1| < \delta_1$. Assume, with no loss of generality that $j' > j$. Then

$$|r^1(t) - r^1(t')| \leq \rho(r^1(t), R(x^1(j\delta_1)))$$

$$+ \sum_{\nu=j}^{j'-1} h(R(x^1(\nu+1)\delta_1)), R(x^1(\nu\,\delta_1)))$$

$$+ \rho(r^1(t'), R(x^1(j'\delta_1)))$$

$$\leq 8\varepsilon_1 + \sum_{\nu=j}^{j'-1} h[R(x^1((\nu+1)\delta_1)), R(x^1(\nu\,\delta_1))]$$

We are now in a position to prove conclusion (a) of theorem 3.1. If R is of bounded variation, it follows (from observation (iii) that $\{r^i(\cdot)\}$ is a bounded equicontinuous sequence in $C[0,T]$. Indeed, given any $\varepsilon > 0$, choose i^* sufficiently large so that $i \geq i^*$ implies $4\varepsilon_i < \varepsilon/2$; now choose $\delta > 0$ such that for $0 \leq \sigma < \delta$, $V_{t-\sigma}^{t} R(x(\cdot)) < \varepsilon/2$ uniformly for $x \in C_b[0,T]$. Then from (iii), above, it follows that if $i \geq i^*$, $|t-t'| < \delta$ then $|r^i(t) - r^i(t')| < \varepsilon$, as desired. Thus $\{r^i\}$ has a uniformly convergent subsequence $\{r^{i_k}(\cdot)\}$ which converges, say, to r. Then $\{x^{i_k}(\cdot)\}$ has a uniformly convergent subsequence, say converging to $x \in C_b[0,T]$. From (ii) we see $r(t) \in R(x(t))$, since $R(x(t))$ is closed, while taking a limit in equation (2) gives $x(t)=x^0+\int_0^t r(\tau)d\tau$. Thus $\dot{x}(t) = r(t) \in R(x(t))$ for all t; $x(0) = x^0$, showing x is a classical solution of (1).

To obtain conclusion (b) of theorem 3.1, if R satisfies property A we find from (iii) that

$$\int_0^T |r^i(t) - r^i(t-\sigma)|dt \leq 4\varepsilon_i T + \int_0^T V_{t-\sigma}^{t}(R(x^i(\cdot)))\, dt.$$

Now given any $\varepsilon > 0$ we can use property A to choose a $\sigma_o > 0$ such that $\int_0^T V_{t-\sigma}^{t} (R(x^i(\cdot)))dt < \varepsilon/2$ if $0 \leq \sigma \leq \sigma_o$, all $i = 1,2,\ldots$. Next choose i^* sufficiently large so that for $i \geq i^*$, $4\,\varepsilon_i T < \varepsilon/2$. Then for $i \geq i^*$, $0 \leq \sigma \leq \sigma_o$, $\int_0^T |r^i(t)-r^i(t-\sigma)|dt < \varepsilon$. By [5, Theorem 20, pp 298], $\{r^i\}$ is $L_1[0,T]$ conditionally compact. Thus it has an $L_1[0,T]$ convergent subsequence, $\{r^{i_k}\}$, which converges, say, to r. Again $\{x^{i_k}\}$ has a uniformly convergent subsequence, convering say, to $x \in C_b[0,T]$. From (ii) we see

$r(t) \varepsilon R(x(t))$ almost everywhere, while taking a limit in equation (2) gives $x(t) = x^o + \int_0^t r(\tau)\, d\tau$ or $\dot{x}(t) = r(t) \varepsilon R(x(t))$ almost everywhere, $x(0) = x^o$, showing that x is a solution (not necessarily a classical solution) of equation (1). This completes the proof of theorem 3.1.

Remark 3.1 : The following question naturally arises. Given an equicontinuous family $\{Q^\alpha : \alpha \varepsilon A\}$ of mappings $Q^\alpha : [0,T] \to \underline{\mathcal{C}}(B^n)$, when does there exist a measurable selection q^α for each Q^α such that the family $\{q^\alpha : \alpha \varepsilon A\}$ is $L_1[0,T]$ conditionally compact. A sufficient condition is shown to be, essentially property (A). It would be interesting, if one could show the existence of such a family $\{q^\alpha : \alpha \varepsilon A\}$ without assuming property (A). This would then lead to an existence theorem for a solution of equation (1) with $R : E^n \to \underline{\mathcal{C}}(B^n)$ merely continuous. The existence of a solution in this case seems to be an open question [4].

Remark 3.2 : We shall show in another section the existence of measurable selections for continuous or lower-semicontinuous mappings $F : X \to 2^Y$ where X is a metric space, Y a complete separable metric space and 2^Y the set of nonempty closed subsets of Y with the Hausdorff topology. In general, one cannot expect more than a Baire selector, indeed if $Q : [0,T] \to \underline{\mathcal{C}}(B^2)$ continuously, many examples exist to show that Q need not admit a continuous selection. Consider the following example due to Hermes [3].

Example 3.1 : For $t \varepsilon [0,1]$ let

$$S(t) = \{ x = \operatorname{Cos} \sigma, \quad y = \operatorname{Sin} \sigma : t \leq \sigma \leq 2\pi \} \subset E^2$$

and for $t > 0$ let

$$A(t) = \begin{pmatrix} \text{Cos } 1/t & \text{Sin } 1/t \\ -\text{Sin } 1/t & \text{Cos } 1/t \end{pmatrix}$$

Define

$$R(t) = \begin{cases} A(t)\, S(t) & \text{if } t > 0 \\ S(0) & \text{if } t = 0 \end{cases}$$

To show R is continuous in the Hausdorff topology, it suffices to show that R is continuous at 0. But $h(R(t), R(0)) \leq \pi t$. If R contains a continuous function r, the graph of r is a connected subset of the cylinder $S(0) \times [0,1]$. But the gap in $R(t)$ for $t > 0$ will disconnect any such arc on $S(0) \times [0,1]$. Thus R is continuous on $[0,1]$ but there does not exist a continuous point valued function r defined on $[0,1]$ with values $r(t) \in R(\varepsilon)$. However one can prove the following theorem [4].

<u>Theorem 3.2</u> : Let $Q : [0,T] \rightarrow \underline{C}(B^n)$ continuously. Suppose Q has bounded variation in $[0,T]$, that is, $V_o^T(Q) < \infty$. Then Q admits a continuous selection r where r is also of bounded variation with same variation as Q.

<u>Proof</u> : For each positive integer k, consider the points $0, T/k, 2T/k, \ldots, T$. Choose $q_o^k \in Q(0)$; $q_1^k \in Q(T/k)$ and such that $|q_o^k - q_1^k| = \rho(q_o^k, Q(T/k))$, and inductively $q_j^k \in Q(jT/k))$. Define $r^k : [0,T] \rightarrow E^n$ as the polygonal arc joining the points q_j^k, $j = 0, \ldots, k$. Then

(i) For any $t \in [0,T]$ and any k, there exists an integer $j = j(k)$ such that $|t - jT/k| < T/k$. Assume with no loss of generality, that $t \in [(j-1)T/k, jT/k]$. Then $\rho(r^k(t), Q(t)) \leq |r^k(t) - r^k(jT/k) + \rho(r^k(jT/k), Q(t)) \leq h(Q(j-1)T/k), Q(jT/k)) + h(Q(jT/k), Q(t))$.

(ii) For $t, t' \in [0,T]$ and any k, let j, j' be integers such that $|t - jT/k| < T/k$, $|t' - j'T/k| < T/k$.

Then $|r^k(t) - r^k(t')| \leq r^k(t) - r^k(jT/k) + \sum_{\nu=j}^{j^1-1} |r^k((\nu+1)T/k) -$

$$r^k(\nu T/k)| + |r^k(j'T/k) - r^k(t')|$$

$$\leq \sum_{\nu=j}^{j'-1} h(Q((\nu+1)T/k), Q(\nu T/k)) + h(Q(t'), Q(jT/k)).$$

We now show that $\{r^k\}$ is equicontinuous. Given any $\varepsilon > 0$, choose k^* sufficiently large so that for $k \geq k^*$, $h(Q(t_1), Q(t_2)) < \varepsilon/3$ if $|t_1 - t_2| < T/k^*$. Next, since Q is of bounded variation, $V_0^t(Q)$ is continuous as a function of t on $[0,T]$, as is $V_{t-\delta}^t(Q)$, hence uniformly continuous, and we can choose a $\delta > 0$ so that $V_{t'}^t(Q) < \varepsilon/3$ if $|t-t'| < \delta$. Then, from (ii), we have : for $k \geq k^*$, $|t-t'| < \delta$, $|r^k(t) - r^k(t')| < \varepsilon/3 + \varepsilon/3 + \varepsilon/3 = \varepsilon$ and equicontinuity is shown. Clearly the sequence $\{r^k\}$ is bounded hence it has a uniformly convergent subsequence, say converging to $r \in C[0, T]$. Let $t \in [0, T]$ and $j(k)$ be an integer such that $|t-j(k)T/k| < T/k$, that is, $j(k)T/k \to t$ as $k \to \infty$. From (i), and the fact that the set $Q(t)$ is closed, it follows that $r(t) \in Q(t)$, that is, r is the desired continuous selection. This completes the proof of theorem 3.2.

Equations of the form (1) arise naturally in many ways. Consider the implicit differential equation $f(t, x, \dot{x}) = 0$, $x(0) = x^0$. This may be reduced to an equation of the form (1) with $R(t,x) = \{ v : f(t, x, v) = 0 \}$. (Note that here $R : [0,1] \times E^n \to E^n$). Another way in which equation (1) appears is in the theory of control systems having equations of the motion of the form $\dot{x} = f(t,x,u)$, $x(0) = x^{o}$, where the control function u may be chosen as any measurable r vector valued function with value at time t in a preassigned set $U(t) \subset E^{r}$. Here

$$R(t,x) = \{ f(t,x,u) : u \in U(t) \}$$

and one may say R has the representation (f, U). We shall study about control problems in another section.

REFERENCES

[1] H. C. Eggleston, Convexity, Cambridge Tracts in Mathematics and Mathematical Physics No. 47, Cambridge University Press London-New York, [1958].

[2] A. F. Filippov, Classical solutions of Differential Equations with multivalued Right Hand Sides (Eng. Trans.) SIAM J. Control, 5 [1967], 609-621.

[3] H. Hermes, The generalised Differential Equation $\dot{x} \in R(t,x)$, Advances in Math 2 [1970], 149-169.

[4] H. Hermes, On continuous and measurable selections and the existence of solutions of Generalised Differential Equations, Proc. Amer. Math. Soc. 29 [1971], 535-542.

[5] N. Dunford, and J. T. Schwartz, Linear Operators I, Interscience Pub. Inc., New York, [1958].

4. MEASURABLE SELECTIONS AND AN APPLICATION TO STOCHASTIC GAMES

The aim of this section is to prove a selection theorem due to Dubins and Savage[7]. We then apply this theorem to prove the existence of optimal stationary strategies for the two players in stochastic games.

We start with some preliminaries. Let Δ be a compact metric space with metric ρ. Denote by 2^{Δ} the collection of all nonempty closed subsets of Δ. We introduce a metric d on 2^{Δ} - the Hausdorff metric (which we defined in section 2), that is, for any $A, B \in 2^{\Delta}$,

$$d(A, B) = \max \left\{ \sup_{x \in A} \rho(x, B), \ \sup_{y \in B} \rho(y, A) \right\}$$

For any sequence $\{ A_n : n = 1, 2, \ldots \}$ define $\overline{\lim} \ A_n = \{ p : p \in \Delta$, there exists an increasing sequence $k_1 < k_2 < \ldots$ of natural numbers such that $p_{k_n} \in A_{k_n}$ and $p_{k_n} \to p \}$. Define $\underline{\lim} \ A_n = \{ p : p \in \Delta, \exists \ p_n \in A_n$ such that $p_n \to p \}$. It is clear that $\overline{\lim} \ A_n$ and $\underline{\lim} \ A_n$ are closed. In case $\overline{\lim} \ A_n = \underline{\lim} \ A_n$ we say that the limit exists and denote it by $\lim A_n$.

Definition : Let X be a metric space and let F be a map from $X \to 2^{\Delta}$. Call F upper semicontinuous in the sense of Kuratowski if $x_n \to x$ implies $\overline{\lim} \ F(x_n) \subset F(x)$.

Now we would like to make the following observations.

(i) $(2^{\Delta}, d)$ is a compact metric space.

(ii) $d(A_n, A) \to 0$ if and only if $\lim A_n = A$.

(iii) If F is u.s.c. from a metric space X into 2^{Δ}, then F is Borel measurable, that is, $\{ x : F(x) \cap B \neq \emptyset \}$ is a Borel subset of X whenever B is a Borel subset of .

See Kuratowski [2] for a proof of these observations.

Now we shall establish the selection theorem of Dubins and Savage. We shall follow the proof as given in Maitra [3]. Let A be a compact metric space, S a Borel subset of a Polish space and v a bounded, upper semicontinuous real valued function on A, that is, $a_n \to a_o$ implies $\limsup v(a_n) \leq v(a_o)$. Assume $|v(a)| \leq M$ for all $a \in A$.

__Lemma 4.1__ : Define $v^* : 2^A \to R$ by $v^*(K) = \max_{a \in K} v(a)$. Then v^* is u.s.c.

__Proof__ : As v is u.s.c. and K compact it follows that there exists $a_o \in K$ such that $v^*(K) = v(a_o)$. Now suppose $K_n \to K$ and assume that for some $a_n \in K_n$, $v^*(K_n) = v(a_n)$. Choose a subsequence $\{ v(a_{n'})$ such that $v(a_{n'}) \to \overline{\lim}\, v^*(K_n)$. As A is compact, there exists a subsequence $\{a_{n''}\}$ of $\{a_{n'}\}$ such that $a_{n''} \to a$. It follows that $a \in K$ and since v is u.s.c. $\limsup v^*(K_n) = \lim v(a_{n''}) \leq v(a) \leq v^*(K)$ which proves v^* is u.s.c.

__Lemma 4.2__ : For each $K \in 2^A$ and $x \in [-M, M]$ define $\tilde{V}(K, x) = \{ a : a \in K \text{ and } v(a) \geq x \}$. Denote by dom $\tilde{V}$ (that is the domain of $\tilde{V}$) the set $\{ (K,x) : (K,x) \in 2^A \times [-M, M]$ and $\tilde{V}(K, x) \neq \emptyset \}$. Then dom $\tilde{V}$ is closed in $2^A \times [-M, M]$ and so a compact metric space. Furthermore $\tilde{V}$ is u.s.c. from dom $\tilde{V}$ to 2^A.

Proof : Since v is u.s.c., for any real c, $\{ a : v(a) \geq c \}$ is an element of 2^A. Next let us show that dom $\tilde{V}$ is closed. Let $(K_n, x_n) \varepsilon$ dom $\tilde{V}$, $n = 1,2,\ldots$, and suppose $(K_n, x_n) \to (K,x)$. Let $a_n \varepsilon \tilde{V}(K_n, x_n)$, $n = 1,2, \ldots$. Since A is compact, there exists a subsequence $\{a_{n_k}\}$ of $\{a_n\}$ such that $a_{n_k} \to a$. Consequently $a \varepsilon K$ and $x = \lim x_{n_k} \leq \limsup v(a_{n_k}) \leq v(a)$, so that $a \varepsilon V(K, x)$. Hence $(K, x) \varepsilon$ dom $\tilde{V}$, which is, therefore closed.

Finally, in order to prove that $\tilde{V}$ is u.s.c. we have to show that $(K_n, x_n) \to (K, x)$, $a_n \varepsilon \tilde{V}(K_n, x_n)$, $n = 1,2,\ldots$, $a_n \to a$ imply that $a \varepsilon \tilde{V}(K, x)$. Since $K_n \to K$, $a \varepsilon K$. Consequently, since $x = \lim x_n \leq \limsup v(a_n) \leq v(a)$, $a \varepsilon \tilde{V}(K, x)$. This completes the proof of lemma 4.2.

Lemma 4.3 : Define V on 2^A by $V(K) = \{a : a \varepsilon K$ and $v(a) = v^*(K)\}$. Then V is a Borel measurable map from 2^A into 2^A.

Proof : As v is u.s.c. $V(K)$ is nonempty. Let us show that it is closed. Let $a_n \varepsilon V(K)$, $n = 1,2,\ldots$, and suppose $a_n \to a$. Then, since v is u.s.c., $v^*(K) = \limsup v(a_n) \leq v(a)$ and as K is closed, $a \varepsilon K$. Consequently $a \varepsilon V(K)$. Hence V maps 2^A into 2^A.

Let D be a Borel subset of 2^A. Note that $V(K) = \tilde{V}(K, v^*(K))$. Consequently,

$$\{ K : K \varepsilon 2^A \text{ and } V(K) \varepsilon D \}$$

$$= \text{proj} [\{(K,x) : (K,x) \varepsilon \text{ dom } \tilde{V} \text{ and } \tilde{V}(K,x) \varepsilon D\} \cap \{(K,x) : v^*(K) = x\}] \quad \ldots (1)$$

As v^* is u.s.c. by lemma 4.1, it is a Borel function from 2^A into $[-M, M]$ and hence its graph is a Borel set in $2^A \times [-M, M]$ (See [2]). Also $\tilde{V}$ is u.s.c. by lemma 4.2, and so it is a Borel map from dom $\tilde{V}$ into 2^A. Consequently,

$$\{(K,x) : (K,x) \in \text{dom } \tilde{V}, \ \tilde{V}(K,x) \in D\}$$

is a Borel subset of dom $\tilde{V}$, which being closed in $2^A \times [-M, M]$, the former set is a Borel subset of $2^A \times [-M, M]$ as well. Thus the set within square brackets on the right-hand side of (1) is a Borel subset of $2^A \times [-M, M]$. Finally, projection being a continuous map, and moreover, 1-1 in this case, it follows by a well-known theorem of Lusin [cf Kuratowski [2], pp 487] that $\{K : K \in 2^A \text{ and } V(K) \in D\}$ is a Borel subset of 2^A. Hence V is a Borel map. This completes the proof of lemma 4.3.

<u>Lemma 4.4</u> : Let u be a bounded u.s.c. function on $S \times A$. Define $u^* : S \to R$ by $u^*(s) = \max_{a \in A} u(s,a)$. Then u^* is u.s.c.

<u>Proof</u> : As u is u.s.c., for fixed s, $u(s,\cdot)$ is u.s.c. in a, so that $u^*(s)$ is well-defined. Let $s_n \to s$ and suppose $u^*(s_n) = u(s_n, a_n)$, $n = 1,2,\ldots$. Choose a subsequence $\{u^*(s_{n'})\}$ such that $u^*(s_{n'}) \to \limsup u^*(s_n)$. Moreover, as A is compact, there is a subsequence $\{a_{n''}\}$ of $\{a_{n'}\}$ such that $a_{n''} \to a$. Since u is u.s.c., it follows that,

$$\limsup u^*(s_n) = \lim_{n''} u(s_{n''}, \ a_{n''}) \le u(s,a) \le u^*(s).$$

Hence u^* is u.s.c. and the proof is complete.

<u>Lemma 4.5</u> : Let u be a bounded u.s.c. on $S \times A$. Define $U : S \to 2^A$ by $U(s) = \{a : a \in A \text{ and } u(s,a) = \max_{a' \in A} u(s,a')\}$. Then U is a Borel map.

Proof of this lemma is similar to lemma 4.3.

Next we shall state and prove the selection theorem due to Dubins and Savage.

Theorem 4.1 : Let u be a bounded u.s.c. on $S \times A$. Then there exists a Borel measurable map f from S into A such that $u(s, f(s)) = \max_{a \varepsilon A} u(s,a)$ for all s. [In this theorem A is assumed to be compact metric and S is a Borel subset of some Polish space].

Proof : Choose a sequence $\{v_n\}$, $n = 1,2,\ldots$ of continuous real valued functions on A, which separate points of A. [For instance one may choose a sequence of functions dense in $C(A)$]. For each v_i, define, for

$$K \varepsilon 2^A, \; v_i(K) = \{ a : a \varepsilon K \text{ and } v_i(a) = \sup_{a' \varepsilon K} v_i(a') \} .$$

Then by lemma 4.3, each V_i a Borel map from 2^A into 2^A. Let U be as in lemma 4.5. Define $U_1(s) = V_1(U(s))$ and $U_n(s)=V_n(U_{n-1}(s))$. By virtue of lemma 4.5 it follows that each U_n is a Borel map from S into 2^A. Moreover, for each s, $U(s) \supset U_1(s) \supset U_2(s)$ ... Since A is compact and $U_i(s)$ are closed in A it follows $\bigcap_{i=1}^{\infty} U_i(s) \neq \emptyset$. Suppose now that for some $s \varepsilon S$, $a, a' \varepsilon \bigcap_{n=1}^{\infty} U_n(s)$ and $a \neq a'$. Then, for every n, as $a,a' \varepsilon U_n(s)$, it follows that $v_n(a) = v_n(a')$, which contradicts the separating property of the sequence $\{v_n\}$. Hence $a = a'$ and for each s, $\bigcap_{n=1}^{\infty} U_n(s)$ is a singleton, say, $\{f(s)\}$.

Next, let us show that for every $s \varepsilon S$, $\{f(s)\} = \lim_{n \to \infty} U_n(s)$ in the Hausdorff metric of 2^A. Fix s and suppose that $a \varepsilon \overline{\lim} U_n(s)$. Then there exists a $\{a_{n_k}\}$ with $a_{n_k} \varepsilon U_{n_k}(s)$ and $a_{n_k} \to a$. As each $U_m(s)$ is closed and $a_{n_k} \varepsilon U_m(s)$ for all $n_k \geq m$ it follows that $a \varepsilon U_m(s)$ and consequently, $a \varepsilon \bigcap_{n=1}^{\infty} U_n(s)$. Hence $\overline{\lim}\, U_n(s) \subset \{f(s)\}$. Also it is clear that $\{f(s)\} \subset \overline{\lim}\, U_n(s)$. Hence $\{ f(s) \} = \lim U_n(s)$. As each U_n is a Borel map from $S \to 2^A$ it now follows that $\Phi : s \to \{f(s)\}$ is a Borel measurable map from S into 2^A. Finally it is not hard to check that the class of all singletons belonging to 2^A is isometric to A. It follows that f is a Borel measurable map from S into A. Moreover as $f(s) \varepsilon U(s)$ for each $s \varepsilon S$, we get $u(s,f(s)) = \max_{a \varepsilon A} u(s,a)$ for every $s \varepsilon S$. This completes the proof of (the selection) theorem 4.1.

Stochastic Games : A stochastic game is determined by five objects : S, A, B, q, r. Here S is a nonempty Borel subset of a Polish space, the set of states of a system; A is a nonempty Borel subset of a Polish space, the set of actions available to Player I ; B is a nonempty Borel subset of a Polish space, the set of actions available to Player II ; q is the law of motion of the system ; it associates (Borel measurably) with each triple $(s,a,b) \varepsilon S \times A \times B$ a probability measure $q(\cdot|s,a,b)$ on the Borel subsets of S ; r, the reward function, is a bounded measurable function on $S \times A \times B$. Periodically (say, once a day) Players I and II observe the current state s of the system and choose actions a and b, respectively ; the choice of the actions is made with full knowledge of the history of the system as it has evolved to the present. As a consequence of the actions chosen by the players, two things happen : Player II pays player I $r(s,a,b)$ units of money, and the system moves to a new state s' according to the distribution $q(\cdot|s,a,b)$. Then, the whole process is repeated from the new state s'. Furthermore, there is specified a discount factor β, $0 \leq \beta < 1$, so that the unit income today is worth β^n at n days in the future. The problem, then, is to maximize the total expected discounted gain of Player I as the game proceeds over the infinite future and to minimize the total expected loss of Player II.

A strategy π for Play I is a sequence $\pi_1, \pi_2, \ldots$, where π_n specifies the action to be chosen by Player I on the nth day by associating (Borel measurably) with each history $h = (s_1, a_1, b_1, \ldots, s_{n-1}, a_{n-1}, b_{n-1}, s_n)$ of the system a probability distribution $\pi_n(\cdot|h)$ on the Borel sets of A. A strategy π for Player I is said to be stationary if there is a Borel map f from S to P_A, where P_A is the set of all probability measures on the Borel sets of A, such that $\pi_n = f$ for each $n \geq 1$; and, in this case, π is denoted by f^∞. Strategies and stationary strategies for Player II are defined analogously.

A pair (π, Γ) of strategies for Players I and II associates with each initial state s an nth-day expected gain $r_n(\pi, \Gamma)(s)$ for Player I and a total expected discounted gain for Player I

$$I(\pi, \Gamma)(s) = \sum_{n=1}^{\infty} \beta^{n-1} r_n(\pi, \Gamma)(s).$$

The functions $r_n(\pi, \Gamma)$ are, plainly, Borel measurable; and consequently, $I(\pi, \Gamma)$ is a Borel function.

A strategy π^* is optimal for Player I if $I(\pi^*, \Gamma')(s) \geq \inf_{\Gamma} \sup_{\pi} I(\pi, \Gamma)(s)$ for all Γ' and all $s \in S$; strategy Γ^* for Player II is optimal if $\sup_{\pi} \inf_{\Gamma} I(\pi, \Gamma)(s) \geq I(\pi', \Gamma^*)(s)$ for all π' and all $s \in S$. We shall say that the stochastic game has a value if $\sup_{\pi} \inf_{\Gamma} I(\pi, \Gamma)(s) = \inf_{\Gamma} \sup_{\pi} I(\pi, \Gamma)(s)$ for every $s \in S$.

In case the stochastic game has a value, the quantity $\sup_{\pi} \inf_{\Gamma} I(\pi, \Gamma)(s)$, as a function on S, is called the value function.

The stochastic game problem was first formulated by Shapley [6] who took S, A, B to be finite, assumed that play would terminate in a finite number of stages with probability one, and considered only what we have called stationary strategies. Shapley was able to prove under these conditions that the stochastic game has a value and that both players have optimal strategies. See [5] or [6]. We shall prove a generalisation of Shapley's result under suitable assumptions on S, A, B, q, r. Specifically, we shall assume that (i) S, A, B are compact metric spaces, (ii) r is a continuous function on $S \times A \times B$, and (iii) whenever $s_n \to s_o$, $a_n \to a_o$, and $b_n \to b_o$, $q(\cdot | s_n, a_n, b_n)$ converges weakly to $q(\cdot | s_o, a_o, b_o)$. These conditions will remain operative throughout the rest of this section.

In order to prove the main theorem, we need some lemmas. If X is a compact metric space, we denote by P_X the space of all probability measures on the Borel sets of X. It is well-known that P_X, endowed with the weak topology, is a compact metric space.

<u>Lemma 4.6</u> : Let i be a continuous, real-valued function on $S \times A \times B$. Then, $i(s, \mu, \lambda) = \iint i(s,a,b) \, d\mu(a) d\lambda(b)$, $s \in S$, $\mu \in P_A$, $\lambda \in P_B$, is a continuous function on $S \times P_A \times P_B$.

For a proof see [4].

<u>Lemma 4.7</u> : Let u be a bounded, continuous function on $X \times Y$ where X is a Borel subset of a Polish space and Y is a compact metric space. Then $u^* : X \to R$ defined by $u^*(x) = \max_{y \in Y} u(x,y)$

is continuous. Moreover $u_* : X \to R$ defined by

$u_*(x) = \min_{y \in Y} u(x,y)$ is continuous.

Proof of this lemma is easy and hence omitted.

<u>Lemma 4.8</u> : Let u be a bounded, continuous function $X \times Y$ where X is a Borel subset of a Polish space and Y is a compact metric space. Then, there exist Borel maps f and g from X into Y such that $u(x, f(x)) = \max_{y \in Y} u(x,y)$, $x \in X$ and $u(x,g(x)) = \min_{y \in Y} u(x,y)$, $x \in X$.

This lemma is an immediate consequence of theorem 4.1.

If X is a topological space, denote by $C(X)$ the family of all bounded, continuous functions on X. For each $w \in C(X)$, define

$$K_w(s,\mu,\lambda) = r(s,\mu,\lambda) + \beta \int w(\cdot)\, dq(\cdot \mid s,\mu,\lambda), \quad s \in S,\ \mu \in P_A,\ \lambda \in P_B$$

where $r(s, \mu, \lambda) = \iint r(s, a, b)\, d\,\mu(a)\, d\,\lambda(b)$ and

$$q(\cdot \mid s, \mu, \lambda) = \iint q(\cdot \mid s, a, b)\, d\,\mu(a)\, d\,\lambda(b).$$

It follows from lemma 4.6, K_w is a continuous function on $S \times P_A \times P_B$. From Sion's minimax theorem [5] we have,

$$\sup_{\mu \in P_A} \inf_{\lambda \in P_B} K_w(s,\mu,\lambda) = \inf_{\lambda \in P_B} \sup_{\mu \in P_A} K_w(s,\mu,\lambda), \quad s \in S.$$

Since K_w is continuous on $S \times P_A \times P_B$ and P_A, P_B are compact, it follows from lemma 4.7 that sup and inf can be replaced above by max and min, respectively. Thus, we have

$$\max_{\mu \in P_A} \min_{\lambda \in P_B} K_w(s,\mu,\lambda) = \min_{\lambda \in P_B} \max_{\mu \in P_A} K_w(s,\mu,\lambda), \quad s \in S.$$

Lemma 4.9 : For each $w \in C(S)$, there exist Borel maps f and g from S into P_A and P_B, respectively, such that

$$\max_{\mu \varepsilon P_A} \min_{\lambda \varepsilon P_B} K_w(s, \mu, \lambda) = \min_{\lambda \varepsilon P_B} K_w(s, f(s), \lambda)$$

and

$$\min_{\lambda \varepsilon P_B} \max_{\mu \varepsilon P_A} K_w(s, \mu, \lambda) = \max_{\mu \varepsilon P_A} K_w(s, \mu, g(s)), \ s \varepsilon S.$$

Proof : We prove the first assertion. Let $\Phi(s,\mu) = \min_{\lambda \varepsilon P_B} K_w(s,\mu,\lambda)$, $s \varepsilon S$, $\mu \varepsilon P_A$. By lemma 4.7, Φ is a continuous function on $S \times P_A$. Hence, by virtue of lemma 4.8, there exists a Borel map $f : S \to P_A$ such that $\Phi(s,f(s)) = \max_{\mu \varepsilon P_A} \Phi(s,\mu) =$ $\max_{\mu \varepsilon P_A} \min_{\lambda \varepsilon P_B} K_w(s, \mu, \lambda)$ for all $s \varepsilon S$. On the other hand, $\Phi(s, f(s)) = \min_{\lambda \varepsilon P_B} K_w(s, f(s), \lambda)$ for all $s \varepsilon S$. This completes the proof of the first assertion.

For each $w \varepsilon C(S)$ define Tw as follows :

$$Tw(s) = \max_{\mu \varepsilon P_A} \min_{\lambda \varepsilon P_B} K_w(s, \mu, \lambda) = \min_{\lambda \varepsilon P_B} \max_{\mu \varepsilon P_A} K_w(s, \mu, \lambda).$$

It is clear from lemma 4.7, T maps $C(S)$ into $C(S)$.

Lemma 4.10 : The operator T is a contraction mapping on $C(S)$.

Proof : Let $w_1, w_2 \varepsilon C(S)$. Plainly, $w_1 \leq w_2 + \| w_1 - w_2 \|$, where, if $w \varepsilon C(S)$, $\| w \|$ denotes $\sup_{s \varepsilon S} |w(s)|$. Since, as is easy to check, T is monotone, we get $Tw_1 \leq T(w_2 + \| w_1 - w_2 \|)$

$$= Tw_2 + \beta \| w_1 - w_2 \| .$$

Consequently $Tw_1 - Tw_2 \leq \beta \|w_1 - w_2\|$. Similarly, (interchanging w_1 and w_2) $Tw_2 - Tw_1 \leq \beta \|w_1 - w_2\|$, which shows $\| Tw_1 - Tw_2 \| \leq \beta \| w_1 - w_2 \|$. Hence, T is a contraction mapping as $\beta < 1$. This completes the proof of the lemma.

Since $C(S)$, when equipped with the supremum norm, is a complete metric space, T has a unique fixed point in $C(S)$, by virtue of the Banach fixed point theorem. Let w^* be the unique fixed point of T. Then it follows from the definition of $K_{w^*}(s, \mu, \lambda)$ and lemma 4.9 that there exist Borel maps f^* and g^* from S to P_A and P_B, respectively, such that, for every $s \in S$,

$$w^*(s) = \min_{\lambda \in P_B} [r(s,f^*(s),\lambda) + \beta \int w^*(\cdot)\, dq\,(\cdot|s,f^*(s),\lambda)]$$

$$= \max_{\mu \in P_A} [r(s,\mu,g^*(s)) + \beta \int w^*(\cdot)\, dq(\cdot|s,\mu,g^*(s))] \quad \ldots(*)$$

$$= r(s,f^*(s), g^*(s)) + \beta \int w^*(\cdot)\, dq(\cdot|s, f^*(s), g^*(s)).$$

We shall show after one lemma that w^* is the value function of the stochastic game.

Our next task is to solve the above functional equation (*). To accomplish this, denote by $M(S)$ the family of all bounded Borel functions on S. With each ordered pair (f,g), where f is a Borel function from S to S to P_A and g is a Borel function from S to P_B, we associate an operator $L(f,g) : M(S) \to M(S)$ defined by

$$(L(f,g)w)(s) = r(s,f(s),g(s)) + \beta \int w(\cdot)\, dq(\cdot|s,f(s),g(s)), \quad s \in S.$$

We may interpret $(L(f,g)w)(s)$ as the expected amount Player II pays Player I, when the initial state of the system is s, Players I and II take actions according to $f(s)$ and $g(s)$, and the game is terminated at the beginning of the second day with Player II paying Player I $w(t)$ units of money, where t is the state of the system on the second day.

Lemma 4.11 : The operator $L(f,g)$ is a contraction mapping on $M(S)$ and $I(f^{\infty}, g^{\infty})$ is its unique fixed point in $M(S)$.

The proof is straight-forward and omit it. Now we are ready to state our theorem [4].

Theorem 4.2 : Let S, A, B be compact metric spaces, let r be a continuous, real-valued function on $S \times A \times B$, and assume moreover, that, whenever $(s_n, a_n, b_n) \to (s_o, a_o, b_o)$ in $S \times A \times B$, $q(\cdot|s_n, a_n, b_n)$ converges weakly to $q(\cdot|s_o, a_o, b_o)$. Then, the stochastic game has a value, the value function is continuous, and Players I and II have optimal stationary strategies.

Proof: Observe that the above functional equation (*) can be rewritten as $L(f^*, g^*) w^* = w^*$. It follows from lemma 4.11 that $w^* = I(f^{*(\infty)}, g^{*(\infty)})$. In view of this we have,

$$I(f^{*(\infty)}, g^{*(\infty)})(s) = \max_{\mu \varepsilon P_A} [r(s,\mu,g^*(s)) + \beta \int I(f^{*(\infty)},g^{*(\infty)})(\cdot)dq(\cdot|s,\mu,g^*(s))]$$

$$= \min_{\lambda \varepsilon P_B} [r(s,f^*(s),\lambda) + \beta \int I(f^{*(\infty)},g^{*(\infty)})(\cdot)dq(\cdot|s,f^*(s),\lambda)]$$

It follows from Blackwell [Theorem 6, pp 232, [1]] or from [Theorem 3.1 in [4]], that,

$$I(f^{*(\infty)},g^{*(\infty)})(s) = \sup_{\pi} I(\pi, g^{*(\infty)})(s)$$

$$= \inf_{\Gamma} I(f^{*(\infty)}, \Gamma)(s) \quad \text{for } s \varepsilon S.$$

Consequently,

$$I(f^{*(\infty)}, g^{*(\infty)}) = \sup_{\pi} I(\pi, g^{*(\infty)}) \geq \inf_{\Gamma} \sup_{\pi} I(\pi, \Gamma).$$

On the other hand,

$$I(f^{*(\infty)}, g^{*(\infty)}) = \inf_{\Gamma} I(f^{*(\infty)}, \Gamma) \leq \sup_{\pi} \inf_{\Gamma} I(\pi, \Gamma).$$

Hence,

$$\inf_{\Gamma} \sup_{\pi} I(\pi, \Gamma) = \sup_{\pi} \inf_{\Gamma} I(\pi, \Gamma).$$

This proves that the stochastic game has a value, that the value function is $I(f^{*(\infty)}, g^{*(\infty)}) = w^*$ and so continuous, and that $f^{*(\infty)}$, $g^{*(\infty)}$ are optimal stationary strategies for Players I and II respectively. This completes the proof of theorem 4.2.

Remark 4.1 : If in theorem 4.2 we allow S to be merely a Borel subset of a Polish space, our proof breaks down because lemma 4.6 fails. However, we can eliminate this difficulty by imposing somewhat stronger conditions on q and r, namely, that $r(s,\mu,\lambda)$ should be a continuous function on $S \times P_A \times P_B$ and that $\int w(.)\, dq\, (.|s, \mu, \lambda)$ should also be continuous on $S \times P_A \times P_B$ for every $w \in C(S)$. [Note that these conditions will be satisfied when A and B are finite]. Then under these conditions on q and r, with A, B compact metric spaces and S a Borel subset of a Polish space, the conclusions and the proof of theorem 4.2 remain valid.

REFERENCES

[1] D. Blackwell, Discounted dynamic programming, Ann. Math. Stat 36 [1965], 226-235.

[2] K. Kuratowski, Topology Vol I, Acad. Press, P.W.N. [1966].

[3] A. Maitra, Discounted dynamic programming on compact metric spaces, Sankhya Series A, 30 [1968], 211-216.

[4] A. Maitra and T. Parthasarathy, On stochastic games, Jour. optimi. theory And its Appl , 5 [1970], 289-300.

[5] T. Parthasarathy and T.E.S. Raghavan, Some topics in two-person games, American Elsevier Publishing Company, New York [1971].

[6] L. S. Shapley, Stochastic games, Proc. National. Acad. Sci U.S.A., 39 [1953], 1095-1100.

[7] L.E. Dubins and L.J. Savage, How to gamble if you must, Mcgraw-Hill, New York [1965].

5. GENERAL THEOREMS ON SELECTORS

In this section we shall first prove a general theorem on selectors due to Kuratowski and Ryll-Nardzewski and deduce a few theorems on selectors [2]. Lastly we shall state a theorem of Jacobs [3].

Let X be a set of arbitrary elements and Y a metric space. Let S be a countably additive family of subsets of X [that is, if $A_n \in S$ for $n = 1,2,\ldots$ then $\bigcup_{n=1}^{\infty} A_n \in S$]. Then the following statement is true.

Lemma 5.1 : Let $f_n : X \to Y$ for $n = 1,2,\ldots,$ and let $f(x) = \lim f_n(x)$ where the convergence is uniform. Suppose that

$$f_n^{-1}(G) \in S \text{ whenever } G \text{ is open in } Y \quad \ldots (1)_n.$$

Then $f^{-1}(G) \in S$ whenever G is open in Y.

Proof : Let $K = \overline{K} \subset Y$. Let $K_n = \{ y : \rho(y, K) \leq 1/n \}$. By assumption there is a sequence $m_1 < m_2 < m_3 < \ldots$ such that $|f_{m_n} - f| \leq 1/n$ for $n = 1,2, \ldots$. We shall show that

$$f^{-1}(K) = \bigcap_{n=1}^{\infty} f_{m_n}^{-1}(K_n).$$

First, let $x \in f^{-1}(K)$, that is, $f(x) \in K$. As $|f_{m_n}(x) - f(x)| \leq 1/n$, we have $f_{m_n}(x) \in K_n$, that is, $x \in f_{m_n}^{-1}(K_n)$ for each n. Let $f_{m_n}(x) \in K_n$, that is, $\rho(f_{m_n}(x), K) \leq 1/n$. As $f(x) = \lim_{n\to\infty} f_{m_n}(x)$, it follows that (owing to the continuity of the metric) that $\rho(f(x), K) = 0$ and hence $f(x) \in \overline{K} = K$. This completes the proof of our lemma.

Let L be a field of subsets of X. [In other words, if A B are members of L, then so are $A \cup B$, $A \cap B$ and $X - A$]. Denote by S the countably additive family induced by L, that is, the family of countable unions of members of L.

Theorem 5.1 [4] : Let Y be a complete separable metric space. Let $F : X \to 2^Y$ (the space of all closed nonvoid subsets of Y) be such that $\{ x : F(x) \cap G \neq \emptyset \} \in S$ whenever G is open in Y. Then there is a selector $f : X \to Y$ such that $f^{-1}(G) \in S$ whenever G is open in Y.

Proof : Let $R = (r_1, r_2, \ldots, r_i, \ldots)$ be a countable set dense in Y. We may suppose of course that the diameter of Y is < 1. We shall define f as the limit of mappings $f_n : X \to R$ where $n = 0, 1, \ldots,$ satisfying condition $(1)_n$ and the two following conditions.

$$(2)_n \quad \ldots \quad \rho(f_n(x), F(x)) < 1/2^n$$

$$(3)_n \quad \ldots \quad |f_n(x) - f_{n-1}(x)| < 1/2^{n-1} \qquad \text{for } n > 0.$$

(The basic idea of proof is similar to the proof of theorem 1.1). Let us proceed by induction. Put $f_o(x) = r_1$ for each $x \in X$. Thus $(1)_o$ and $(2)_o$ are fulfilled. Now let us assume, for a given $n > 0$, that f_{n-1} satisfies conditions $(1)_{n-1}$ and $(2)_{n-1}$. Put

$$C_i^n = \{ x : \rho(r_i, F(x)) < 1/2^n \},$$

$$D_i^n = \{ x : |r_i - f_{n-1}(x)| < 1/2^{n-1} \}, \quad \text{and}$$

$$A_i^n = C_i^n \cap D_i^n .$$

We have, $X = A_1^n \cup A_2^n \cup \ldots$. For, x being a given point of X, there is by $(2)_{n-1}$, $y \in F(x)$ such that $|y - f_{n-1}(x)| \leq 1/2^{n-1}$. Since $\{r_1, r_2, \ldots\}$ is dense, we can find a r_i such that $|r_i - y| < 1/2^n$ and $|r_i - f_{n-1}(x)| < 1/2^{n-1}$. Hence $x \in A_i^n$.

Denote by B_i^n the open ball $\{y : |y - r_i| < 1/2^n\}$. It follows that $C_i^n = \{x : F(x) \cap B_i^n \neq \emptyset\}$ and $D_i^n = f_{n-1}^{-1}(B_i^{n-1})$. Hence it follows that $C_i^n \in S$ and $D_i^n \in S$ and consequently $A_i^n \in S$. Consequently $A_i^n = \bigcup_{j=1}^{\infty} E_{i,j}^n$, where $E_{i,j}^n \in L$. Arrange the double sequence (i,j) in a simple sequence (k_s, m_s) where $s = 1,2,\ldots,$ and put

$$E_s^n = E_{k_s,m_s}^n .$$

We have, $X = E_1^n \cup E_2^n \cup \ldots E_s^n \cup \ldots$

This identity allows us to define a mapping $f_n : X \to R$ as follows: $f_n(x) = r_{k_s}$ if $x \in E_s^n - (E_1^n \cup \ldots \cup E_{s-1}^n)$. We shall show that f_n satisfies $(1)_n$, $(2)_n$ and $(3)_n$. By definition $f_n^{-1}(r_{k_s}) = E_s^n - (E_1^n \cup \ldots \cup E_{s-1}^n)$. As L is a field, it follows that $f_n^{-1}(r_{k_s}) \in L$ and as $f_n^{-1}(r_i) = \bigcup_{k_s=i} f_n^{-1}(r_{k_s})$, we have $f_n^{-1}(r_i) \in S$ for each i. Consequently $f_n^{-1}(Z) \in S$ for each $Z \subset R$ (since R is countable and S countably additive). Thus $(1)_n$ is satisfied.

For a given x let s satisfy,

$$x \in E_s^n - (E_1^n \cup \ldots \cup E_{s-1}^n).$$

Put $k_s = i$. Hence we have $x \in E_s^n \subset A_i^n = C_i^n \cap D_i^n$ and it is clear that f_n's satisfy $(2)_n$ and $(3)_n$. Thus the sequence

$f_o, f_1, \ldots, f_n \ldots$ has been defined according to the conditions $(1)_n$, $(2)_n$ and $(3)_n$.

By $(3)_n$ and by the completeness of the space Y, this sequence converges uniformly to a mapping $f : X \to Y$. By lemma 5.1, it follows $f^{-1}(G) \in S$ whenever G is open in Y. Finally $f(x) \in F(x)$ according to $(2)_n$. Thus the proof of theorem 5.1 is complete.

<u>Remark 5.1</u> : The theorem remains true by replacing the condition [namely $\{ x : F(x) \cap G \neq \emptyset \} \in S$ whenever G is open in Y] by $\{ x : F(x) \cap K \neq \emptyset \} \in L$ whenever $K \subset Y$ is closed. This can be seen as follows. Y being metric, every open $G \subset Y$ is an F_σ set : $G = K_1 \cup K_2 \cup \ldots$, where $K_n = \overline{K}_n$. Hence

$$\{ x : F(x) \cap G \neq \emptyset \} = \bigcup_{n=1}^{\infty} \{ x : F(x) \cap K_n \neq \emptyset \} \in S.$$

<u>Remark 5.2</u> : For each complete separable metric space Y, there is a choice function $f : 2^Y \to Y$ of the first class of Baire. Also f may be assumed to be continuous if $\dim Y = 0$ (that is, if Y contains a countable base composed of closed-open sets). These observations can be seen as follows. Put in theorem 5.1, $X = 2^Y$, L = field of subsets of 2^Y which are F_σ and G_δ and F = the identity mapping defined on 2^Y. According to the Vietoris topology, the sets $\{ K : K \cap G \neq \emptyset \}$ are open and the sets $\{ K : K \cap Q \neq \emptyset \}$ are closed in 2^Y provided G is open and Q is closed in Y. As G is F_σ in Y, so $\{ K : K \cap G \neq \emptyset \}$ is F_σ in 2^Y, hence a member of L. It is not difficult to check (from theorem 5.1) that there is a choice function $f : 2^Y \to Y$ of the first class of Baire, that is, $f^{-1}(G)$ is an F_σ set for

every open G in Y. In the particular case where $\dim Y = 0$, we denote by L the field of closed-open subsets of 2^Y. By assumption we have $G = Q_1 \cup Q_2 \cup \ldots$, where Q_n is closed open. Consequently $\{ K : K \cap Q_n \neq \emptyset \} \in L$ and $\{ K : K \cap G \neq \emptyset \}$ belong to S. Since the members of S are open sets, it follows that $f^{-1}(G)$ is open for every open set G in Y and hence f is continuous.

Now we shall deduce the following two theorems from theorem 5.1.

Theorem 5.2 [2] : Let X be a set with σ-ring S and let Y be a T_2-space which is the union of a family of at most Ω (= first uncountable ordinal) compact metrizable subspaces in such a way that any compact subset of Y lies in the union of an utmost countable subfamily. Let $F : X \to 2^Y$ (closed subsets of Y). Suppose

$$F^{-1}(C) = \{ x : F(x) \cap C \neq \emptyset \} \in S \quad \text{whenever } C \text{ is compact in } Y.$$

Then there exists an $f : X \to Y$ such that $f(x) \in F(x)$ for every $x \in X$ and $f^{-1}(C) \in S$ whenever C is compact in Y.

Definition : A Lusin space is a separable metrizable space which is the image of a complete separable metric space under a continuous 1-1 function.

Theorem 5.3 : Let X be a set with σ-algebra S and Y be a Lusin space. Let $F : X \to 2^Y$ (= nonempty closed subsets of Y). Suppose

$$\{ x : F(x) \cap B \neq \emptyset \} \in S \quad \text{for every Borel set } B \text{ in } Y.$$

Then there exists an $f : X \to Y$ such that $f(x) \in F(x)$ for every x and $f^{-1}(B) \in S$ for every Borel set B in Y.

Proof of theorem 5.2 : Let the family of compact sets described in the hypothesis be $\{ Y_\alpha : \alpha < w \}$ where w is an ordinal less than or equal to Ω. For each $\alpha < w$, define $X_\alpha = F^{-1}(Y_\alpha) - \cup\{ F^{-1}(Y_\beta) : \beta < \alpha \}$. Clearly, the X_α's are pairwise disjoint, and their union is X. Assign to each X_α the σ-ring S_α obtained by restricting S to X_α. Each S_α is, in fact, a σ-algebra on X_α (that is $X_\alpha \in S_\alpha$), since X_α is the intersection of the countable family $\{ F^{-1}(Y_\alpha) - F^{-1}(Y_\beta) : \beta < \alpha \}$ and each of the sets $F^{-1}(Y_\beta)$, $\beta \leq \alpha$, is a member of S. Moreover, since $X_\alpha \in S$, we have that $S_\alpha \subset S$ for all α.

Note that, by the definition of X_α, we have that $F(x) \cap Y_\alpha \neq \emptyset$ for all $\alpha < w$, $x \in X_\alpha$. Thus for all $\alpha \in w$, define a point-closed function $F_\alpha : X_\alpha \rightarrow 2^{Y_\alpha}$ by $F_\alpha(x) = F(x) \cap Y_\alpha$ if $x \in X_\alpha$. It is clear that $F_\alpha^{-1}(C) = X_\alpha \cap F^{-1}(C) \in S_\alpha$ for each closed (and therefore compact) subset C of Y_α. Further Y_α is complete with any compatible metric. Thus by theorem 5.1, for each α we have an f_α such that $f_\alpha(x) \in F_\alpha(x)$ and $f_\alpha^{-1}(C) \in S_\alpha$ for every closed subset C of Y_α. Now define $f : X \rightarrow Y$ by

$$f(x) = f_\alpha(x) \quad \text{if} \quad x \in X_\alpha, \quad \alpha < w.$$

Clearly, f is a selector for F. Furthermore, $f^{-1}(C) \in S$ for every compact subset C of Y. For, let C be a compact subset of Y, and let α be an at most countable ordinal such that

$$C \subset \cup\{ Y_\beta : \beta \leq \alpha \}.$$

Then $\beta > \alpha$ implies that,

$$f_\beta^{-1}(C \cap Y_\beta) \subset F_\beta^{-1}(C) \subset \cup\{ F_\beta^{-1}(Y_r) : r \leq \alpha \} = \emptyset,$$

since $r \leq \alpha < \beta$ implies that

$$F_\beta^{-1}(Y_r) = \{ x \in X_\beta : F_\beta(x) \cap Y_r \neq \emptyset \}$$
$$\subset \{ x \in X_\beta : F(x) \cap Y_r \neq \emptyset \} = X_\beta \cap F^{-1}(Y_r) = \emptyset .$$

It follows that $f^{-1}(C) = \cup \{ f_\beta^{-1}(C \cap Y_\beta) : \beta \leq \alpha \}$. Plainly $f_\beta^{-1}(C \cap Y_\beta) \in S_\beta \subset S$ for all $\beta \leq \alpha$. Hence $f^{-1}(C) \in S$ and thus the proof is complete.

<u>Proof of theorem 5.2</u> : Let $\phi : P \to Y$ be a 1-1 continuous function from a complete separable metric space P onto Y. Define $F_* : X \to P$ by $F_* = \phi^{-1} \circ F$. $F_*(x)$ is closed for each $x \in X$, since ϕ is continuous. Moreover $\{ x : F_*(x) \cap G \neq \emptyset \} \in S$ for every G open in P. For let G be open in P. Then G has a compatible metric which makes it a complete metric space. Hence, the subspace $\phi(G)$ of Y is a Lusin space. Hence we can conclude that $\phi(G)$ is a Borel subset of Y. It follows that $F_*^{-1}(G) = F^{-1}(\phi(G)) \in S$. Applying theorem 5.1, we have an $f_* : X \to P$ such that $f_*(x) \in F_*(x)$ and $f_*^{-1}(G) \in S$ for every G open in P. Define $f = \phi \circ f_* : X \to Y$. Then f is a selector for F since

$$f(x) = \phi(f_*(x)) \in \phi(F_*(x)) = \phi \circ \phi^{-1}(F(x)) = F(x)$$

for every $x \in X$. To prove that $f^{-1}(B) \in S$ for every Borel set B in Y, it is sufficient to prove $f^{-1}(B) \in S$ for every open set B in Y. Thus, suppose that B is open in Y. Then $\phi^{-1}(B)$ is open by the continuity of ϕ, and $f^{-1}(B) = f_*^{-1}(\phi^{-1}(B)) \in S$ and thus the proof of the theorem is complete.

Remark 5.3 : Theorem 5.3 generalises a theorem of Castaing [1] where he assumes X to be a compact metric space and S the corresponding family of Borel sets.

We shall end this section by stating one more selection theorem due to Jacobs [3]. For this we need some terminology. Let X be a complete separable metric space. Let μ denote a positive Radon measure defined on a compact T_2 space T and let Y be a metric space. Let $f : T \times X \to Y$ with $f(t,\cdot)$ locally uniformly continuous in x for every fixed t and $f(\cdot, x)$ measurable in t for every fixed x. Let $y : T \to Y$ be a measurable mapping such that $y(t) \in f(t,X)$ for every t and $\Gamma : T \to 2^X$ denotes the mapping defined by

$$\Gamma(t) = \{ x : x \in X \text{ and } f(t,x) = y(t) \}$$

Now we are set to state our final theorem of this section.

Theorem 5.4 : Under the assumptions stated in the above paragraph, there exists a measurable function $x : T \to X$ such that $x(t) \in \Gamma(t)$ for each $t \in T$. For a proof refer to Jacobs [3].

REFERENCES

[1] C. Castaing, Quelques problems de measurabilite lies a la theorie de la commande, C.R. Acad. Sci. Paris 262 [1966], 409-411.

[2] C. J. Himmelberg and F. S. Van Vleck, Some selection theorems for measurable functions, Canadian. Jour. Math 21 [1969], 394-399.

[3] Marc. Q. Jacobs, Remarks on some recent extentions of Filippov's implicit functions lemma, SIAM. J. Control 5 [1967], 622-627.

[4] K. Kuratowski and C. Ryll-Nardzewski, A general theorem on selectors, Bulletin De L'aAcademi Polonaise Des Sciences (Serie des sciences math. astr. et phys) 13 [1965], 397-403.

6. TWO APPLICATIONS OF MEASURABLE SELECTIONS

In this section we shall give two applications of measurable selection theorems (which we have proved in section 5). The first result is an example of a non-analytic (in fact non-Lebesgue measurable) subset of $[0,1]$ which is a Blackwell space, due to Michael Orkin [3]. The second result is a theorem on stochastic games [4].

We will start with some preliminaries.

<u>Definition</u> : A countably generated σ-algebra $\mathfrak{F}$ of subsets of a set X is called a Blackwell space if every countably generated σ-algebra $\mathcal{C} \subset \mathfrak{F}$ having the same atoms as $\mathfrak{F}$ coincides with $\mathfrak{F}$, that is, $\mathcal{C} = \mathfrak{F}$.

In [1] Blackwell proved that every analytic subset of a Polish space (endowed with the relative Borel σ-field) is a Blackwell space. We shall now construct an example of a non-analytic subset of $[0,1]$ which is a Blackwell space.

<u>The construction</u> : Let $I = [0,1]$. Let $\mathcal{B}$ be the σ-field of Borel sets of I. Using transfinite induction, we will construct a set $A \subset I$ with the following properties :

a1) Neither A nor A^c contains an uncountable analytic set.

a2) If f is a countable to one Borel function from $I \to I$ such that $S_f = \{ x : f(x) \neq x \}$ is uncountable then there exists $x \in A$ such that $f(x) \in A$ and $f(x) \neq x$.

We proceed with the construction. We first well-order the class of uncountable Borel sets in I. (This class has power c). We next well-order the class of all Borel functions which satisfy the following conditions.

f 1) f is countable to one, that is $\{f^{-1}(x)\}$ is at most countable for every x.

f 2) S_f is uncountable where $S_f = \{x : f(x) \neq x\}$.

This class of functions also has power c. We inductively construct two disjoint collections of nested sets A_α, E_α as follows.

We select distinct points $x_1, y_1 \in B_1$ (the first uncountable Borel set in our ordering) and distinct points $r_1, s_1 \in I - \{x_1 - y_1\}$ such that $f_1(r_1) = s_1$. We let $A_1 = \{x_1, r_1, s_1\}$, $E_1 = \{y_1\}$. [Here f_1 is the first function in our ordering. When B_α, f_α, are reached in our inductive procedure (where α is an ordinal $< c$), we select x_α, y_α from $B_\alpha - \bigcup_{\beta<\alpha} (A_\beta \cup E_\beta)$. We can do this because every uncountable Borel set has power c. We also select distinct points r_α, s_α from $I - \{\bigcup_{\beta<\alpha} (A_\beta \cup E_\beta) \cup \{x_\alpha, y_\alpha\}\}$ such that $f_\alpha(r_\alpha) = s_\alpha$. We can do this by observing that the set of points we have thus far removed has power $< c$, and by noting the conditions imposed on f_α, including the fact that S_{f_α}, being uncountable and Borel, must also have power c. We then let

$$A_\alpha = \{\cup (A_\beta : \beta < \alpha)\} \cup \{x_\alpha, r_\alpha, s_\alpha\} \quad \text{and}$$

$$E_\alpha = \{\cup (E_\beta : \beta < \alpha)\} \cup \{y_\alpha\}$$

We then let $A = \bigcup_{\alpha < c} A_\alpha$.

By the method of construction, A, A^c (complement of A) do not contain uncountable Borel sets, hence no uncountable analytic sets, so condition (a 1) is satisfied. Also any Borel function f satisfying (f 1) and (f 2) appeared somewhere in the induction process so (a 2) is satisfied.

We now consider the pair $(A, \mathcal{B}^*)$ where $\mathcal{B}^*$ is the relative Borel σ-field on A, that is, $\mathcal{B}^*$ = sets of the form $B \cap A$, $B \in \mathcal{B}$.

Theorem 6.1 : If $\mathcal{F}^*$ is a countably generated subfield of $\mathcal{B}^*$(on A) whose atoms (on A) are the points of A, then $\mathcal{F}^* = \mathcal{B}^*$. [Here we assume the continuum Hypothesis].

We need the following selection theorem due to E. A. Stchegolkow.

Proposition 6.1 : Let $f : Z \to W$ be a Borel measurable countable to one function where Z and W are Borel subsets of some Polish spaces. Then there exists a Borel subset E of Z such that (i) $f|E$ is 1-1 and (ii) $f(E)$ = range of f.

Proof: Note that if f is a countable to one continuous function defined on a complete separable metric space taking values in a Polish space then the forward image of every Borel set is also Borel - for a proof of this see Kuratowski, Topology Vol 1.

It is enough to prove proposition 6.1 when f is continuous, countable to one and $Z = N^N$ (= the set of all sequences of positive integers.

Let X = range of f. Define $F: X \to 2^Z$ (nonempty closed subsets of Z), such that, $F(x) = f^{-1}(x)$.

Plainly $F(x)$ is closed for each x since f is continuous. F is also measurable since, the set,

$$\{ x : F(x) \cap U \neq \emptyset \} = f(U)$$

is Borel (by the remark made above) for every open set U in Z. Hence by Karatowski — Ryll-Nardzewski's theorem (5.1). F has a selection g. That is, $g : X \to Z$ is Borel measurable and

$g(x) \in F(x) = f^{-1}(x)$ for every x. Plainly g is 1-1. Take $E = g(X)$. Then E has all required properties stated in proposition 6.1.

Proof of theorem 6.1 : Let $F_n = B_n \cap A$ be a countable generating class for $\mathfrak{F}^*$ and consider the sub-σ-field of $\mathcal{B}$ on I generated by the sets B_n. Call this σ-field $\mathfrak{F}$. We claim the atoms of are countable sets.

If T is an atom of $\mathfrak{F}$, since $\mathfrak{F}$ separates points on A, T contains at most one point in A. Since T is a Borel set ($\mathfrak{F}$ being countably generated) and because T contains at most one point of A, $T \cap A^c$ is Borel. Thus $T \cap A^c$ is a countable set for A^c contains no uncountable Borel set. Thus T is a countable set.

We now use the Borel selection theorem, that is proposition 6.1. This proposition implies the existence of a Borel function g from $I \to I$ which maps each atom of $\mathfrak{F}$ onto a point in itself. Since the atoms of $\mathfrak{F}$ are countable sets, g must be countable to one. [See the remark 6.1 below]. We claim $S_g = \{ x : g(x) \neq x \}$ is countable. Indeed if S_g is uncountable, then g belongs to the class of functions we used in the construction of A, and there must exist distinct points $x, y \in A$ with $g(x) = y$. This is impossible since g maps each atom of $\mathfrak{F}$ into itself and no atom of $\mathfrak{F}$ contains more than one A point. But the fact that S_g is countable implies that all but a countable number of the atoms of $\mathfrak{F}$ are singletons. Thus $\mathfrak{F}$ separates the points of I with the exception of at most a countable set. From this it follows that $\mathfrak{F}^* = \mathcal{B}^*$. Thus the proof of theorem 6.1 is complete.

Remark 6.1 : Our use of the selection theorem was accomplished by using the well-known fact that if $\mathfrak{F}$ is a countably generated σ-field on I, say, then there is a Borel function f which maps distinct atoms into distinct points (not necessarily in the same atom) and from which $\mathfrak{F}$ can be got by mapping the Borel sets of I through f^{-1} (for example, if $F_1, F_2, \ldots,$ generate $\mathfrak{F}$, let

$$f = \sum_{n=1}^{\infty} \frac{2}{3^n} \cdot f_n \quad \text{where} \quad f_n(x) = 1 \text{ if } x \in F_n \text{ and } = 0 \text{ if } x \notin F_n).$$

If we consider the graph of f, which is a Borel set in the corresponding product space, the horizontal sections of the graph correspond to the atoms of $\mathfrak{F}$. Since in the case, we were considering, these sections were then countable sets, the conditions for the selection theorem were satisfied and the selection function so obtained mapped each atom of $\mathfrak{F}$ onto a point in itself.

Before stating our next result on stochastic games, we shall write down our assumptions. [See section 4 for details regarding stochastic games]. S will be a complete separable metric space. A and B are finite sets r(s, a, b) is a bounded measurable function of s for fixed a, b. Now we shall state our theorem [4].

Theorem 6.2 : Under the assumptions stated above, the stochastic game has a value and the value function is Borel measurable. Furthermore Players I and II have optimal stationary strategies.

Remark 6.2 : Let $r_w : S \to P_A \times P_B$ where

$$r_w(s) = \{ (\mu',\lambda') : \max_{\mu} [r(s,\mu,\lambda') + \beta \int w(\cdot)\, dq(\cdot\,|s,\mu,\lambda')]$$

$$= \min_{\lambda} [r(s,\mu',\lambda) + \beta \int w(\cdot)\, dq(\cdot\,|s,\mu',\lambda)\}$$

and w is a bounded real-valued Borel measurable function on S. These set valued functions r_w are measurable by a result of Olech [2]. Now one can use (the selection) theorem 5.3 to prove theorem 6.2. The rest of the proof follows along similar lines to that of theorem 4.2, with some minor modifications and hence the proof of theorem 6.2 is omitted.

REFERENCES

[1] D. Blackwell, On a class of probability spaces, Proceedings of the third Berkeley Symposium on Mathematical Statistics and Probability Vol 2, p 1-6, edited by J. Neyman; University of California Press [1956].

[2] C. Olech, A note concerning set valued measurable functions, Bulletin De L'Academie Polonaise Des Sciences, Serie des Sciences, math, astro et Phys. 13 [1965], 317-321.

[3] M. Orkin, A Blackwell space which is not analytic, To appear.

[4] T. Parthasarathy, Discounted and positive stochastic games, Bull. Amer. Math. Soc 77 [1971], 134-136.

7. VON-NEUMANN'S MEASURABLE CHOICE THEOREM

The purpose of this section is to prove a measurable choice theorem due to von-Neumann. After giving the proof, we shall mention a few consequences of this theorem.

Choice Theorem 7.1 [4] : Let S be a complete separable metric space, A an analytic set in S and $F(a)$ a function which is defined and continuous for all $a \varepsilon A$ and has real numerical values. Let K be the set of all values, which $F(a)$, $a \varepsilon A$ assumes. Let $\tau(\lambda)$ be an arbitrary N-function, that is, $\tau(\lambda)$ is a monotone, nondecreasing, right-continuous, bounded function. Under these assumptions we have

(i) K is a $\tau(\lambda)$ measurable set.

(ii) there exists a function $f(\lambda)$, which is defined for all $\lambda \varepsilon K$ and assumes values in S such that

(a) If O is an open subset of S, then the set of all $\lambda \varepsilon K$ with $f(\lambda) \varepsilon O$ is $\tau(\lambda)$-measurable.

(b) For every $\lambda \varepsilon K$, $F(f(\lambda)) = \lambda$.

Proof : We shall prove the theorem when $\tau(\lambda)$ is a continuous and nowhere constant N-function, therefore $\lambda \to \tau(\lambda) = \mu$ is a topological mapping of $-\infty < \lambda < \infty$ on the finite interval $a < \mu < b$ where $a = \tau(-\infty)$, $b = \tau(+\infty)$. The general situation can be reduced to this case [4].

Let us now replace $F(a)$, $f(\lambda)$ by $(F(a))$, $f(\tau^{-1}(\mu))$, and then write again λ for μ. This leaves the structure of our assertions unaffected, but there are these changes regarding λ. The domain of λ is now $a < \lambda < b$ and not $-\infty < \lambda < +\infty$.

In this new domain we have to use the notion of common (Legesgue)-measurability and not that one of $\tau(\lambda)$-measurability. [Our mapping $\lambda \to \tau(\lambda)$ transformed the latter into the former].

We now introduce the '0 space of Baire', S_o, which is the set of all sequences $\alpha = (n_1, n_2, \ldots)$, $n_1, n_2 \ldots = 1, 2, \ldots$. It is metrized by the definition

$$\text{Dist}\,[(m_1, m_2, \ldots), (n_1, n_2, \ldots)] = 0 \quad \text{if } m_v = n_v \text{ for every } v.$$
$$= \frac{1}{v_o} \quad \text{if } v_o \text{ is the smallest integer for which } m_v \neq n_v.$$

Thus S_o is a complete, separable metric space. [The countable set of all those $\alpha = (n_1, n_2, \ldots)$, for which $n_v \neq 1$ occurs for a finite number of v's only is obviously everywhere dense in S_o]. On the other hand the definition, $(m_1, m_2, \ldots) < (n_1, n_2, \ldots)$ if a v_o exists such that $m_{v_o} < n_{v_o}$ and $m_v = n_v$ for $v < v_o$, establishes an ordering of S_o. If α is fixed, then the set of all $\beta < \alpha$ is open. If $\beta < \alpha$, and v_o is defined as above, then we have $\delta < \alpha$ in the entire sphere $\text{Dist}\,(\delta, \beta) < 1/v_o$. Every sphere in S_o is a set of the form $\alpha_1 \leq \beta < \alpha_2$ (with fixed α_1, α_2). If its center is $(m_1, m_2, \ldots)$ and its radius $\varepsilon > 0$, then $\beta = (n_1, n_2, \ldots)$ belongs to it, if and only if, $\text{Dist}\,\{(m_1, m_2, \ldots), (n_1, n_2, \ldots)\} < \varepsilon$, that is $m_v = n_v$ for $v = 1, 2, \ldots, v_1$ where v_1 is the greatest integer $\leq 1/\varepsilon$. This may be written in the form $\alpha_1 \leq \beta < \alpha_2$ if we put

$$\alpha_1 = (m_1, \ldots, m_{v_1-1}, m_{v_1}, 1, 1, \ldots), \quad \alpha_2 = (m_1, \ldots, m_{v_1-1}, m_{v_1}+1, 1, 1, \ldots).$$

As A is an analytic set in a complete, separable metric space, therefore there exists a function $\Phi(\alpha)$ which is defined and continuous for every $\alpha \in S_0$ and which maps S_0 onto A. [The mapping may be many to one]. There K is clearly the set of all $F(\Phi(\alpha))$, $\alpha \in S_0$. As $F(\Phi(\alpha))$ also is defined and continuous for every $\alpha \in S_0$ and the real numbers form a complete, separable metric space, we can conclude that K is an analytic set and consequently measurable.

Consider now a $\lambda \in K$. Denote by the set of all $\alpha \in S_0$ with $F(\Phi(\alpha)) = \lambda$ by $T(\lambda)$. As $\lambda \in K$, $T(\lambda)$ is not empty and as $F(\Phi(\alpha))$ is continuous, $T(\lambda)$ is closed. Denote the smallest n_1 with $(n_1, n_2, \ldots) \in T(\lambda)$ for any $(n_2, n_3, \ldots)$ by n_1^0, the smallest n_2 with $(n_1^0, n_2, n_3 \ldots) \in T(\lambda)$ for any $(n_3, n_4, \ldots)$ by n_2^0 etc. Let $\alpha(\lambda) = (n_1^0, n_2^0, n_3^0, \ldots)$. It is clearly a condensation point of $T(\lambda)$ and therefore $\alpha(\lambda) \in T(\lambda)$, besides its definition guarantees that $\alpha(\lambda) \leq \beta$ for every $\beta \in T(\lambda)$. In other words $T(\lambda)$ has a first element, and this is $\alpha(\lambda)$.

Choose now a fixed α_0 and consider the set $M(\alpha_0)$ for all λ's with $\alpha(\lambda) < \alpha_0$. That is, $\lambda \in M(\alpha_0)$ is clearly equivalent to the existence of a $\beta < \alpha_0$ with $\beta \in T(\lambda)$, that is of a $\beta < \alpha_0$ with $F(\Phi(\beta)) = \lambda$. In other words $M(\alpha_0)$ is the image of open set $\{\beta : \beta < \alpha_0\}$ in S_0 under $F \circ \Phi$. Hence we can conclude $M(\alpha_0)$ is an analytic set and hence measurable.

If T is any subset of S_0, denote the set of all λ's with $\alpha(\lambda) \in T$ by $N(T)$. If T is a sphere, then it has the form $\alpha_1 \leq \beta < \alpha_2$, therefore $N(T) = M(\alpha_2) - M(\alpha_1)$. Our above result proves therefore, that $N(T)$ is measurable. If T is merely an open set, then, as S_0 is separable, we can write T as the sum of a sequence of spheres : $T = \bigcup_{i=1}^{\infty} T_i$. Now $N(T) = \bigcup_{i=1}^{\infty} N(T_i)$, and as $\{N(T_i)\}$ are all measurable, so is $N(T)$.

Let finally O be an open subset of S, and denote by $T(O)$ the set of all $\alpha \in S_0$ with $\phi(\alpha) \in O$. As O is open and $\phi(\alpha)$ is continuous, therefore $T(O)$ is open too. Thus $N(T(O))$ is measurable. That is, $\lambda \in N(T(O))$ means $\alpha(\lambda) \in T(O)$ and hence $\phi(\alpha(\lambda)) \in O$.

Observe finally, that as $\alpha(\lambda) \in T(\lambda)$ (assuming $\lambda \in K$), $F(\phi(\alpha(\lambda))) = \lambda$.

The measurability of K being already established, we have proved statement (i) of our theorem. And in order to prove statement (ii) it suffices to define $f(\lambda) = \phi(\alpha(\lambda))$. Thus the proof of theorem 7.1 is complete.

For the statement of our next theorem we need some definitions. We shall need the concept of standard measurable space. This is a measurable space that is isomorphic to a cartesian product of $\{0,1\}$ countably many times. The space $\{0,1\}$ consists of the points 0 and 1 only and all subsets are measurable. An isomorphism between two measurable spaces is a one-one function from one onto the other, that takes measurable sets onto measurable sets in both directions. The unit interval, any Euclidean space, and in fact any uncountable Borel subset of any separable complete metric space, with the usual Borel σ-field, is standard. Now we are ready to state a theorem due to Aumann [1].

<u>Theorem 7.2</u> : Let (T, μ) be a σ-finite measure space, let X be a standard measurable space, and let G be a measurable subset of $T \times X$ whose projection on T is all of T. Then there is a (Lebesgue) measurable function $g : T \to X$, such that $(t, g(t)) \in G$ for almost all t in T. ['Almost all' means that the exceptional set is measurable and of μ-measure 0].

<u>Remark 7.1</u>: This theorem follows from the previous theorem — for a proof refer [1]. Also, the theorem is false without the requirement that X be standard, as the following example due to Lindenstrauss shows. Let T be the half open interval $[0,1)$ with the Borel sets, and let μ be Lebesgue measure. Let X be the subspace of $[0,1)$ that one usually uses to show that there are non-Lebesgue measurable sets; that is, the sets $X + r$ are mutually disjoint for rational r, and $\bigcup_r (X + r) = [0,1)$ (addition is modulo 1). Let D be the subset of $T \times X$ defined by

$$D = \{ (t, x) : t = x \} .$$

Here D is a sort of diagonal of $T \times X$. It is not hard to check that D is measurable ; one builds a finite number of small rectangles whose union covers D, lets the rectangles get smaller, and the intersection is D. Note that the projection of D on X exhausts X, but the projection on T does not exhaust T ; in fact, the projection is precisely X, which may be considered a subset of T.

Let $$G = \bigcup_r (D + (r, 0)).$$

Clearly $D + (r, 0)$ is measurable, and hence G is measurable. The projection of $D + (r, 0)$ on T is $X + r$, and hence the projection of G exhausts T. Since the $X + r$ are disjoint, it follows that for each t there is precisely one point $g(t) \varepsilon X$ such that $(t,g(t)) \varepsilon G$. Clearly it is sufficient to prove that there is no measurable function that differs from g on a set of Lebesgue measure 0 only.

We first prove a lemma: If S is a measurable subset of T such that $S + r = S$ for all rational r, then $\mu(S) = 0$ or 1. To prove this suppose $\mu(S) > 0$. Then we can find a point t_o at which S has density 1; hence for every ε and every sufficiently small interval $[s,t)$ around t_o, we will have $\mu(S \cap [s,t)) > (1-\varepsilon)(t-s)$. In particular, if m is a sufficiently large positive integer, we can find an interval $[s,t)$ such that $t - s = 1/m$ and $\mu(S \cap [s,t)) > (1-\varepsilon)/m$. Then

$$\begin{aligned}
\mu(S) &= \mu(S \cap [0,1)) = \mu(S \cap \sum_{j=1}^{m} ([s,t) + j/m)) \\
&= \sum_{j=1}^{m} \mu(S \cap ([s,t) + j/m)) \\
&= \sum_{j=1}^{m} \mu((S + j/m) \cap ([s,t) + j/m)) \\
&= \sum_{j=1}^{m} \mu((S \cap [s,t)) + j/m) \\
&= \sum_{j=1}^{m} \mu(S \cap [s,t)) > m(1-\varepsilon)/m = 1 - \varepsilon.
\end{aligned}$$

Since ε may be chosen arbitrarily small, the lemma follows. From this lemma we deduce the following Corollary. If S is a not necessarily measurable subset of $[0,1)$ such that $S + r = S$ for all rational r, then the outer measure of S is either 0 or 1. Indeed, let S' be a measurable set containing S, such that $\mu(S') = \mu^+(S)$, where μ^+ denotes outer measure. (For example, take S' to be the intersection of a sequence $\{S_i\}$ of measurable sets containing S, such that the measure of S_i differs from $\mu^+(S)$ by $1/i$). Then $S' + r$ is measurable for all r, and includes $S + r = S$. Therefore $S'' = \cap S' + r$ is measurable, and includes S; since $S'' \subset S'$, it follows that $\mu(S'') = \mu^+(S)$. Now $S'' + r = S''$ for all rational r, hence $\mu(S'') = 0$ or 1 and the Corollary is proved.

Obviously the outer measure of X cannot vanish; let Y be a measurable subset of T such that $Y \supset X$ and $\mu(Y) = \mu^+(X) > 0$. Using density considerations as in the proof of the lemma, we can find two disjoint intervals $I_1 = [s_1, t_1)$ and $I_2 = [s_2, t_2)$ such that $\mu(Y \cap I_1) > 0$, $\mu(Y \cap I_2) > 0$. Then also $\mu^+(X \cap I_1) > 0$, $\mu^+(X \cap I_2) > 0$. Let

$$X_1 = \bigcup_r ((X \cap I_1) + r)$$

$$X_2 = \bigcup_r ((X \cap I_2) + r).$$

Then $X_1 + r_1 = X_1$ and $X_2 + r = X_2$ for all rational r. But $X_1 \supset X \cap I_1$, and $\mu^+(X_1) \geq \mu^+(X \cap I_1) > 0$; similarly $\mu^+(X_2) > 0$. So from the Corollary it follows that $\mu^+(X_1) = 1$, $\mu^+(X_2) = 1$, and since $X_1 \cap X_2 = \emptyset$, it follows that $\mu^-(X_1) = \mu^-(X_2) = 0$, where μ^- denotes the inner measure. But $X_1 = g^{-1}(I_1)$, and it follows that there is no measurable function g' such that g differs from g' on a set of measure zero. This completes the example.

We shall present another version of von-Neumann's choice theorem. Let X be an arbitrary complete separable metric space. The function F will be defined on $T = [0,1]$ but its values will be subsets of X. Recall a point valued function f from T to X will be called Lebesgue measurable if $f^{-1}(U)$ is a Lebesgue measurable subset of T for every open (or equivalently, Borel) subset of X. An analytic subset of X is the continuous image of a Borel subset of X. The set-valued function F will be called analytic if its graph, that is,

$$\{ (t, u) : u \in F(t) \}$$

is an analytic subset of $T \times X$. Now we have the following proposition [2].

Proposition 7.1 : If F is an analytic set-valued function from T to X, then there is a Lebesgue measurable point-valued function $f : T \to X$ such that $f(t) \in F(t)$ for almost all t.

Proof follows from theorem 7.1.

Next we give an application of proposition 7.1 in the characterisation of extreme points of sets of vector functions [2,3].

Let A be a compact convex subset of E^n, B the set of its extreme points, $\bar{B}$ the closure of B. Let M_A, M_B and $M_{\bar{B}}$ be the sets of all measurable functions from $T = [0,1]$ to A, B and $\bar{B}$ respectively. Since the space of all measurable functions from T to E^n has a linear structure, it is meaningful to talk about the extreme points of M_A.

Proposition 7.2 : The set of extreme points of M_A is precisely M_B.

<u>Proof</u> : Clearly every point of M_B is an extreme point of M_A. Conversely, let f be an extreme point of M_A. If f is not in M_B, then for some t, $f(t)$ is not an extreme point of A. Hence for each t we may choose $g(t)$ and $h(t)$ in A so that $f(t) = \frac{1}{2} g(t) + \frac{1}{2} h(t)$, and $g(t)$ and $h(t)$ differ from $f(t)$ for at least some t. Because of proposition 7.1, g and h may be chosen so as to be measurable as well. Then g and h are in M_A and $f = \frac{1}{2} g + \frac{1}{2} h$, contradicting the extreme property of f. This proves the proposition.

Another situation treated by Karlin [3] is the following. Let $\mu_1, \mu_2, \ldots, \mu_m$ be a set of non-atomic totally finite measures on T, and let $a_1, \ldots, a_m$ and $b_1, \ldots, b_m$ be in E^n. Let A and B be as above. Let G M_A, consiting of functions f such that $b_i \leq \int f \, d\mu_i \leq a_i$ for $i = 1, 2, \ldots, m$. Karlin proved that the extreme points of G are contained in $M_{\bar{B}}$. We shall now prove the following proposition.

<u>Proposition 7.3</u> : The extreme points of G are contained in M_B.

<u>Proof</u> : We follow the proof due to Aumann [2]. Let f be an extreme point of G and suppose $f \notin M_B$. Construct g and h as in the previous proof, and let $e = \frac{1}{2} g - \frac{1}{2} h$ so that $f + e = g \in M_A$ and $f - e = h \in M_A$. For $S \subset T$ set $\mu(S) = \{ \int_S e \, d\mu_1, \ldots, \int_S e \, d\mu_m \}$. Then μ is a vector measure of dimension m. Applying Lyapunov's theorem on vector measures, we obtain a subset S of T such that $\mu(S) = \frac{1}{2} \mu(T)$. Define e' on T by $e'(t) = e(t)$ for $t \in S$, and $e'(t) = -e(t)$ for $t \notin S$. Define $f_1 = f + e'$, $f_2 = f - e'$. Then

$f = \frac{f_1 + f_2}{2}$, $f_1, f_2 \varepsilon M_A$ and

$$\int f_1 \, d\mu_1 = \int f \, d\mu_1 + \int_S e \, d\mu_1 - \int_{T/S} e \, d\mu_1$$

$$= \int f \, d\mu_1 + \frac{1}{2} \int e \, d\mu_1 - \frac{1}{2} \int e \, d\mu_1 = \int f \, d\mu_1 .$$

Hence $f_1 \varepsilon G$. Similarly $f_2 \varepsilon G$ and the proof is complete.

REFERENCES

[1] R. J. Aumann, Measurable utility and the measurable choice theorem, Proc. Int. Colloq., La Decision, C.N.R.S, Aix - en - Provence [1967], 15-26.

[2] R. J. Aumann, Integrals of set-valued functions, Jour. Math. Analy. Appl. 12 [1965], 1-12.

[3] S. Karlin, Extreme points of vector functions, Proc. Amer. Math. Soc. 4 [1953], 603-610.

[4] J. von Neumann, On rings of operators, Reduction theory, Ann. Math. 50 [1949], pp 448-451.

8. ON THE UNIFORMISATION OF SETS IN TOPOLOGICAL SPACES

Given a set E in the cartesian product $X \times Y$ of two spaces X and Y, a set U is said to uniformise E, if the projections $\pi_X E$ and $\pi_X U$ of E and U through Y onto X coincide, and if, for each $x \in \pi_X E$ the set

$$\{(x) \times Y\} \cap U \qquad \ldots (1)$$

of points of U lying above x consists of a single point. The existence of such a uniformising set U follows immediately from the axiom of choice. But, if X and Y are topological spaces and E is, in some sense topologically respectable, for example if E belongs to some Borel class, it is natural to ask a uniformising set U that is equally respectable or at any rate not much worse. Usually there is no way of controlling the topological respectability of sets obtained by the use of axiom of choice and quite different methods have to be used in obtaining topologically respectable uniformising sets.

The earlier work of Lusin [3] on problems of this nature was confined to the case when for each x in $\pi_X E$, the set (1) of points of E lying above x is at most countable. The first general result seems to have been the result obtained independently by Lusin [4] and Sierpinski [7] showing that, when X and Y are Euclidean spaces and E is a Borel set in $X \times Y$, the uniformising set U can be taken to be the complement of an analytic set. Following work by Lusin and Novikoff [5] on the effective choice of a point from a complement of an analytic set defined by a given sieve, Kondo [6] showed that in the Euclidean case, the complement of an analytic set could be uniformised by a complement of an analytic set. Since any Borel set in a Euclidean space is the complement of an analytic set this provided a most satisfactory generalisation of the result of Lusin and Sierpinski. Braun [2] showed in the Euclidean plane that any closed set E can be uniformised by a G_δ set and that any F_σ set can be uniformised by a $G_{\delta\sigma}$ set.

It is easy to extend the results of Lusin and Sierpinski and of Kondo when X and Y are complete separable metric spaces by the following mapping technique. If X is a complete separable metric space there will be a continuous function f that maps a relatively closed subset I_0 of the set I of irrational numbers in $[0,1]$ regarded as a subset of R_1, one-one onto X. Similarly there will be a continuous function g that maps a relatively closed subset J_0 of the set J of irrational numbers in $[0,1]$ regarded as a subset of the set S_1 of real numbers, one-one onto Y. Then the product map $f \times g$ maps $I_0 \times J_0$ continuously and one-one onto $X \times Y$. Hence the inverse map $f^{-1} \times g^{-1}$ maps Borel sets and coanalytic sets in $X \times Y$ into Borel sets and coanalytic sets in $I_0 \times J_0$, which remain Borel sets and coanalytic sets in $R_1 \times S_1$. So the results of Lusin and Sierpinski or of Kondo can be applied in $R_1 \times S_1$; when the uniformising set is intersected with $I_0 \times J_0$ and mapped by $f \times g$ back to $X \times Y$ there results in $X \times Y$ a uniformising set that is the complement of an analytic set as required. We shall now prove the following theorem due to Rogers and Willmott [6].

<u>Theorem 8.1</u> : Let X be any topological space. Let Y be any σ-compact metric space. Let $\mathcal{D}$ be the family of finite unions of differences of closed sets in $X \times Y$. Then a closed set in $X \times Y$ can be uniformised by a $\mathcal{D}_\delta$ set and a F_σ set in $X \times Y$ can be uniformised by a $\mathcal{D}_{\delta\sigma}$-set.

We will now introduce some notations and definitions. When we study a cartesian product $X \times Y$ of two spaces we use π_X, π_Y to denote the projection operators onto X and Y respectively, so that

$$\pi_X(x \times y) = x, \quad \pi_Y(x \times y) = y \quad \text{for } x \in X,\ y \in Y.$$

If C is any set in $X \times Y$ we use $C^{(x)}$, for $x \in X$, to denote the set of y in Y with $x \times y \in C$ and $C^{(y)}$ is defined similarly. We also define the cylinder on a set E to be the set

$$\pi_X E \times Y$$

and use $\text{cy}\, E$ to denote the cylinder.

A set U will be said to uniformise a set E in $X \times Y$ if $U \subset E$, $\pi_X U = \pi_X E$ and $U^{(x)}$ contains a single point for each x in $\pi_X E$.

A function f is said to uniformise a set E in $X \times Y$ if f is defined on $\pi_X E$ into Y so that

$$x \times f(x) \in E \quad \text{for every } x \in \pi_X E.$$

We will call a set Z in $X \times Y$ a cylinder parallel to Y if it is of the form $Z = P \times Y$ for some subset P of X.

<u>Lemma 8.1</u> : Let Y^* be a compact subset of a T_2 space Y. Let J be a set of the form $F - Z$ in $X \times Y$ with F closed and Z a closed cylinder parallel to Y. Then the sets

$$J_o = J \cap (X \times Y^*)$$

$$J_1 = J - \text{cy } J_o$$

are both of the same form as J and

$$\pi_X J_o \cap \pi_X J_1 = \emptyset$$

$$\pi_X J_o \cup \pi_X J_1 = \pi_X J.$$

<u>Proof</u> : Let $J = F - Z$ with F closed and Z a closed cylinder parallel to Y. Then

$$J_o = \{ F - Z \} \cap \{ X \times Y^* \} = \{ F \cap (X \times Y^*) \} - Z$$

is of the required form, as $F \cap (X \times Y^*)$ is closed. As Y^* is compact, it follows easily that $\pi_X \{ F \cap (X \times Y^*) \}$ is a closed set in X. Hence the set

$$Z_1 = \text{cy } [F \cap (X \times Y^*)] = [\pi_X \{ F \cap (X \times Y^*) \}] \times Y$$

is a closed cylinder parallel to Y. Further

$$\text{cy } J_o = \text{cy } [\{F \cap (X \times Y^*)\} - Z] = Z_1 - Z$$

So $J_1 = J - \text{cy } J_o = \{ F - Z \} - \{ Z_1 - Z \}$

$$= F - (Z \cup Z_1)$$

and J_1 has the required form.

The formulae for the intersection and union of the projections of J_o and J_1 follow immediately from the facts that J_o is a subset of J and that $J_1 = J - cy\, J_o$. This proves the lemma.

<u>Proof of theorem 8.1</u> : Let E be a closed in $X \times Y$. As Y is σ-compact and metric, we can choose a sequence $Y_1, Y_2 \ldots$ of compact subsets of Y with the properties :

(a) the diameter of Y_i tends to zero as $i \to \infty$ and

(b) each point of Y belongs to infinitely many sets of the sequence.

We define sets $D_o, D_1, D_2, \ldots$ inductively by taking $D_o = E$,

$$D_{n+1} = \{ D_n \cap (X \times Y_{n+1})\} \cup \{ D_n - cy(D_n \cap (X \times Y_{n+1}))\} \qquad \ldots (2)$$

for $n = 0, 1, 2, \ldots$. It follows inductively, by the use of the lemma that D_n is the union of 2^n sets, each of the form $F - Z$ with F closed and Z a closed cylinder parallel to Y, having disjoint projections with union $\pi_X E$.

Write $U = \bigcap_{n=0}^{\infty} D_n$.

Then each set $D_n \in \mathfrak{D}$ and U is a $\mathfrak{D}_\delta$ set. As

$$\pi_X U \qquad \pi_X D_o = \pi_X E$$

it will suffice to show that given any $x \in \pi_X E$, there is a unique point y_x with $x \times y_x \in U$.

If $x \in \pi_X E$, the set $D_o^{(x)} = E^{(x)}$ is closed and nonempty. Our aim is to show that the sets $D_n^{(x)}$, $n = 0,1,2, \ldots$, are closed, decreasing, nonempty, compact for n sufficiently large and with diameter tending to zero. This will ensure that the set $U^{(x)} = \bigcap_{n=o}^{\infty} D_n^{(x)}$ consists of a single point as required.

It follows from (2) that,

$$\begin{aligned} D_{n+1}^{(x)} &= D_n^{(x)} \cap Y_{n+1} \quad \text{if} \quad D_n^{(x)} \cap Y_{n+1} \neq \emptyset \\ &= D_n^{(x)} \quad \text{if} \quad D_n^{(x)} \cap Y_{n+1} = \emptyset \end{aligned}$$

for $n = 0,1,2, \ldots$. It follows immediately that the sets $D_n^{(x)}$, $n = 0,1,2, \ldots$, are closed, decreasing and non-empty. As the sequence $Y_1, Y_2, \ldots$ covers Y, it follows that for some first integer $n(x)$, $D_{n(x)}^{(x)} \cap Y_{n(x)+1} \neq \emptyset$. This implies that $D_n^{(x)}$ is compact for $n \geq n(x) + 1$. As each point of Y lies in infinitely many sets of the sequence $Y_1, Y_2, \ldots$, and as the diameters of these sets tend to zero, it follows that the diameter of $D_n^{(x)}$ also $\to 0$. This completes the proof of the case when E is closed.

Now consider the case when E is an F_σ set. As Y is σ-compact we can express E in the form $E = \bigcup_{n=1}^{\infty} E_n$, where each set E_n is closed and each set $\pi_Y E_n$ is a subset of a compact set in Y. Then each set $\pi_X E_n$ is closed (being effectively the projection of a closed set through a compact space). Let U_n be a D_δ set uniformising E_n, for $n = 1, 2, \ldots$.

Let $U_n = \bigcap_{m=1}^{\infty} D_{nm}$ where each set $D_{nm} \varepsilon \mathcal{D}$. It is clear that the set

$$U = \bigcup_{n=1}^{\infty} [U_n - cy (\bigcup_{\nu < n} U)]$$

uniformises E. But

$$U_n - cy(\bigcup_{\nu < n} U) = \bigcap_{m=1}^{\infty} \{D_{nm} - cy (\bigcup_{\nu < n} E)\}$$

As π_X E is closed for $\nu < n$, the set

$$D_{nm} - cy (\bigcup_{\nu < n} E)$$

belongs to $\mathcal{D}$. Hence U is a $\mathcal{D}_{\delta\sigma}$-set as required.

<u>Remark 8.1</u> : If each open set in X is a F_σ set, the uniformising set for a closed set can be taken to be a G_δ set and that for a F_σ set can be taken to be a $G_{\delta\sigma}$-set. This can be seen as follows. In this case each open set in $X \times Y$ is a countable union of open rectangles $U \times V$. Then, as each open set U in X and V in X and V in Y is an F_σ-set, each open set in $X \times Y$ is an F_σ-set. So each closed set and hence each set in $\mathcal{S}$ is a G_δ set in $X \times Y$. The result follows.

We shall close this section after stating one more result due to Rogers and Willmott, a proof of which can be found in [6].

<u>Theorem 8.2</u> : Suppose that each open set in a topological space X is a Souslin set. Let E be the complement of a Souslin set

in $X \times I$ where $I = [0,1]$. Then there is a set U, that is the complement of a Souslin set in $X \times I$ and that satisfies,

(a) $U \subset E$

(b) $\pi_X U = \pi_X E$ and

(c) for each $x \in \pi_X E$, the set $(\{x\} \times I) \cap U$ contains a single point.

Remark 8.2 : It seems that Robert Solovay has proved the following result [See [8], Remark 4, pp 2] assuming the existence of measurable cardinals. Every complement of an analytic set in two-dimensional Euclidean plane can be uniformised by a Lebesgue measurable set ! See also pp 24-25, of [1] in this connexion.

REFERENCES

[1] R. J. Aumann, Measurable utility and the measurable choice theorem, Proc. Int. Colloq., La Decision, C.N.R.S., Aix-en-Provence [1967], 15-26.

[2] S. Braun, Sur l'uniformisation des ensembles fermes, Fund. Math, 28 [1937], 214-218.

[3] N. Lusin, Lecons sur les ensembles analytiques et leurs applications. Paris, [1930].

[4] N. Lusin, Sur le probleme de M.J. Hadamard d'uniformisation des ensembles. Mathematica, 4 [1930], 54-66.

[5] N. Lusin and P. Novikoff, Choix effectif d'un point dans un complementaire analytic arbitraire, donne par un crible, Fund. Math, 25 [1935], 559-560.

[6] C. A. Rogers and R. C. Willmott, On the uniformisation of sets in topological spaces, Acta. Math 120 [1968], 1-52.

[7] W. Sierpinski, Sur l'uniformisation des ensembles measurables (B), Fund. Math 16, [1930], 136-139.

[8] R. Solovay, A model of set theorey in which every set of reals is Lebesgue measurable, Ann. Math 92, [1970], pp 2.

9. SUPPLEMENTARY REMARKS ON SELECTION THEOREMS

The purpose of this section is to make a few remarks on selection theorems and to mention applications which we have not covered so far. We may not give detailed proof to any of the theorems to be stated in this section but we give references where one can find the complete proof.

(a) <u>n-selections</u> : Suppose X and Y are metric spaces. Let Φ be a function from $X \to 2^Y$, sometimes called a carrier. (Here 2^Y denotes the set of closed subsets of Y). A function σ from X to 2^Y is said to be an at most n-selection for Φ if for each $x \in X$, $\sigma(x)$ is a subset of $\Phi(x)$ containing at most n points and $\bigcup_{x \in X} \sigma(x)$ is closed in $\bigcup_{x \in X} \Phi(x)$. In order to relate a topology of 2^Y to X or the topology of Y to that of X when $\Phi : X \to 2^Y$ is a lower semicontinuous function, it is convenient to require that Φ have the following properties.

<u>Definition</u> : A carrier $\Phi : X \to 2^Y$ is said to be point-compact if the set L_y (defined below)

$$L_y = \{ x : y \in \Phi(x) \}$$

is compact for each $y \in Y$.

A carrier will be called small if for each $y \in Y$ and open set O, $L_y \subset O \subset X$, there is an open neighbourhood V of y for which

$$L_y \subset \{x : \Phi(x) \cap V \neq \emptyset\} \subset O.$$

Remark 9.1 : If $f : Y \to X$ is an open and onto mapping, then $\Phi : X \to 2^Y$ defined by $\Phi(x) = f^{-1}(x)$ is a lower semicontinuous, small and point compact carrier. Also if $\Phi : X \to 2^Y$ is upper semicontinuous, X is compact and Y is regular then Φ is small and point-compact. The following theorem has been proved in [6] by McAuley and Addis.

Theorem 9.1 : Suppose X and Y are metric spaces with $\dim X \leq n$ and that $\Phi : X \to 2^Y$ is a lower semicontinuous, small and point-compact carrier for which $\Phi(x)$ is complete with respect to the metric on Y for every $x \in X$. Furthermore, suppose that $B \subset X$ is a closed set (perhaps empty) and $g : B \to Y$ is a map so that $g(x) \in \Phi(x)$ for each $x \in B$. Then there is an at most (n+1)-selection σ for Φ such that $\sigma(x) = g(x)$ for $x \in B$.

Remark 9.2 : The proof depends on the fact that certain projections are closed mappings (which is not true in general) but our assumptions on Φ ensures that they are closed.

(b) Continuous carriers : We will now briefly mention about continuous carriers discussed in [8]. If Y is a metric space (metric ρ), then a carrier $\Phi : X \to 2^Y$ is continuous if, given $\varepsilon > 0$, every $x_0 \in X$ has a neighbourhood U such that, for every $x \in U$,

$$\Phi(x_0) \subset N_\varepsilon(\Phi(x)), \qquad \Phi(x) \subset N_\varepsilon(\Phi(x_0))$$

where $N_\varepsilon(\Phi(x)) = \{ y : \rho(y, \Phi(x)) < \varepsilon \}$ etc. It is clear that Φ is continuous in this sense, if and only if it is continuous, in the usual sense, with respect to the following topology τ on 2^Y. This τ is the topology generated by the uniform structure $\mathcal{U}$ on 2^Y which is obtained by taking, as a basis for the entourages of the diagonal in $2^Y \times 2^Y$, all classes of the form

$$\mathcal{V}_\varepsilon = \{(A,B) : A,B \in 2^Y, \quad A \subset N_\varepsilon(B), \quad B \subset N_\varepsilon(A)\},$$

with $\varepsilon > 0$. Now one can prove the following results. If X is a topological space and Y a Banach space then every continuous $\Phi : X \to \mathfrak{F}(Y)$ (= nonempty, closed, convex subsets of Y) admits a selection. If X is a topological space, and Y normed linear space, then every continuous carrier ψ from X to the nonempty open, convex subsets of Y admits a selection. In these two results no restriction is placed on X except that it is a topological space. It would be nice if one could prove a selection theorem when Y is an arbitrary metric - of course one has to have strong assumptions on the map Φ as well as the sets $\Phi(x)$. For example if X is finite dimensional (here we mean Lebesgue covering dimension, that is, dim $X \leq n$ if every finite open covering $\mathcal{U}$ of X has a finite open refinement $\mathcal{V}$ such that every $x \in X$ belongs to at most $n + 1$ elements of $\mathcal{V}$), it is possible to place purely topological conditions on the sets $\Phi(x)$ which are not only sufficient but, in a sense, also necessary. See [7] in this connexion.

(c) Part metric in convex sets : Now we will briefly discuss some results on the part metric and its relation to selection theorem due to Bauer and Bear. We will start with some preliminaries.

Consider a real linear space L, and a convex set C in L which contains no whole line. The closed segment from x to y is denoted by $[x, y]$. If $x, y \in C$ we say that $[x,y]$ extends in C by $r(>0)$ if $x + r(x-y) \in C$ and $y + r(y-x) \in C$. We write $x \sim y$ if $[x,y]$ extends by some $r > 0$. It can be shown that $\sim$ defines an equivalence relation in C. The equivalence classes of $\sim$, called the parts of C, are clearly also convex. There is a metric on each part of C defined by

$$d(x,y) = \inf \{ \log(1 + \frac{1}{r}) : [x,y] \text{ extends by } r \}.$$

If x is not equivalent to y, write $d(x, y) = +\infty$. Then d satisfies all the axioms of a metric on C, except that it is not always finite. This metric d will be called the part metric on C.

<u>Definition</u> : A metric d on C will be called convex if

(i) the mapping $(\lambda, x, y) \to \lambda x + (1-\lambda)y$ of $[0,1] \times C \times C$ into C is continuous and

(ii) the set $\{ y : y \in C, d(y, S) < \varepsilon \}$ is convex for each $\varepsilon > 0$ and each convex subset S of C.

The part metric of a convex set C without any line is convex on each part π of C – for a proof see [2]. Now we are in a position to state the following theorem.

<u>Theorem 9.2</u> : Let Φ be a lower semicontinuous mapping from a paracompact space T to the nonempty closed convex subsets of one part π of the cone $M_+(X)$ of nonnegative Radon measures on a locally compact space X. Then there exists a continuous

selection for Φ. Let f be any such function and $\mu \varepsilon \pi$. Then f has the form $f(t) = g_t \mu$, where $t \to g_t$ is a continuous mapping from T into $L^\infty(\mu)$ (= space of μ-essentially bounded μ-measurable functions].

The proof of this theorem depends on the following observation. The part metric d in each part π_μ (containing μ) is topologically equivalent to the $L^\infty(\mu)$-metric on the corresponding functions $P_\mu = \{ g : g \varepsilon L^\infty(\mu),\ g$ is positive and bounded away from $0 \}$.

For a proof see pp 27 in [2]. Compare this theorem with Michael's theorem in section 1. Here ''the Banach space Y'' is replaced by a convex set with a complete metric which is convex in the sense described above.

(d) Fixed point theorem and the selection problem : Kakutani has proved the following result. Let M be a compact convex subset of the Euclidean n-space and let T be an upper semi-continuous set valued function from M into itself such that for every $x \varepsilon M$, the image set $T(x)$ is nonempty, closed and convex. Then T admits a fixed point, that is, there exists a point x_o with $x_o \varepsilon T(x_o)$. We will presently show how this theorem can be viewed as a selection problem. For this we need some preliminiaries.

Let $S = [a_1, a_2, \dots a_{n+1}]$ stand for an n-simplex with vertices $a_1, a_2, \dots, a_{n+1}$. Given a simplex $[a_1, a_2, \dots, a_j]$, the point $\frac{1}{j}[a_1 + a_2 + \dots + a_j]$ is called the barycentre of this simplex. The barycentres of all the faces of orders 0, 1, 2, ... of a simplex S determine a family $S^{(1)}$ of n simplexes, called the barycentric division of order 1 of S, with the properties :

(i) each n simplex of $S^{(1)}$ contains the barycentre of a face of dimension 0, of a face of dimension 1, etc.

(ii) if an n simplex of $S^{(1)}$ contains the barycentres of $[a_{j_1}, a_{j_2}, \ldots a_{j_k}]$ and $[a_{i_1}, a_{i_2}, \ldots, a_{i_p}]$, where $k < p$, then $\{ j_1, j_2, \ldots, j_k \} \subset \{ i_1, i_2, \ldots, i_p \}$.

If we divide each n-simplex of $S^{(1)}$ in a similar manner, we obtain a new family of n-simplexes, called the barycentric division of order 2, which we denote by $S^{(2)}$. Continuing this process, we can define the barycentric division $S^{(k)}$ of order k.

Let $T : S \to S$ be an upper semicontinuous set valued function such that for each $x \in S$, $T(x)$ is a nonempty closed subset of S where S is an n-simplex. For each vertex a^k of the kth barycentric division $S^{(k)}$ of the simplex S, let b^k be a point of $T(a^k)$ and write $b^k = \Phi_k(a^k)$. For each $x \in S$, we consider a n-simplex $[a^k_{i_1}, a^k_{i_2}, \ldots, a^k_{i_{n+1}}] \in S^{(k)}$ which contains it and we put

$$\Phi_k(x) = \Phi_k(p_1 a^k_{i_1} + p_2 a^k_{i_2} + \ldots) = p_1 \Phi_k(a^k_{i_1}) + p_2 \Phi_k(a^k_{i_2}) + \ldots .$$

The function Φ_k so defined is linear in the interior of a simplex of $S^{(k)}$ and is therefore continuous ; moreover it is uniquely determined, even at the points which belong to several n-simplexes of $S^{(k)}$; consequently Φ_k is continuous in S. Hence by

Brouwer's fixed point theorem (which asserts that every single-valued continuous function from S into itself has a fixed point), there exists a point $x_k \in S$ such that $x_k = \Phi_k(x_k)$. Since S is compact, the sequence $\{x_k\}$ has a cluster point x_o. If $[a_1^k, a_2^k, \ldots a_{n+1}^k]$ is an n-simplex of $S^{(k)}$ which contains the point x_k, we can write

$$x_k = \Phi_k(x_k) = \Phi_k\,(p_1^k\, a_1^k + p_2^k\, a_2^k + \ldots) = p_1^k\, \Phi_k(a_1^k) + p_2^k\, \Phi_k(a_2^k) + \ldots$$

$$= p_1^k\, b_1^k + p_2^k\, b_2^k + \ldots + p_{n+1}^k\, b_{n+1}^k$$

where $b_1^k \in T(a_1^k)$ and $b_2^k \in T(a_2^k)$, etc.

Without any loss of generality we shall assume that,

$x_k \to x_o$, $b_i^k \to b_i^o$, $p_i^k \to p_i^o$ as $k \to \infty$ for $i = 1,2,\ldots n+1$.

Since $a_i^k \to x_o$, $b_i^k \to b_i^o$, $b_i^k \in T\, a_i^k$ and T is upper semi-continuous, we have $b_i^o \in T\, x_o$ for $i = 1,2, \ldots n+1$.

Now we have the following proposition.

<u>Proposition 9. 1</u> : Let S be an n-simplex. Let Φ_k be functions defined as above and $T : S \to S$ be upper semi-continuous with $T(x)$ nonempty and closed for each $x \in S$. Then given a $\varepsilon > 0$ there exists an integer N such that for all $k \geq N$,

$$d(\Phi_k(x),\ T(x)) < \varepsilon\ .$$

<u>Remark 9.3</u> : If the single-valued functions Φ_k (as defined above) converges uniformly to a single-valued function Φ_o then

Φ_o is a selection for T. This can be seen as follows. From the definition of the functions Φ_k and the uniform convergence of the functions Φ_k to Φ_o, we have,

$$d(\Phi_o(x), T(x)) \leq d(\Phi_o(x), \Phi_k(x)) + d(\Phi_k(x), T(x)) \leq \varepsilon + \varepsilon = 2\varepsilon$$

for $n > N(\varepsilon)$. Now we are ready to state the following theorem (which is a slight extention of Kakutani's fixed point theorem) due to Patnaik [9].

Theorem 9.3 : Let T be an upper semicontinuous map from $S \rightarrow S$ with T(x) nonempty and closed. Let Φ_k converge uniformly. Then there exists a fixed point under T.

The truth of this theorem can be seen as follows. We have $\Phi_k(x_k) = x_k$ and $x_k \rightarrow x_o$ as $k \rightarrow \infty$. Since $\Phi_k \rightarrow \Phi_o$ uniformly and since Φ_o is a selection for T, it follows that $x_o \in T(x_o)$. In Kakutani's theorem we assume each T(x) is nonempty convex and closed but here T(x) need not be convex for each x. Of course here we assume Φ_k converges uniformly to ensure the existence of a fixed point.

(e) Control problems of contingent equation : Now we shall consider the control problem and sketch a proof due to Kikuchi for the existence of optimal control [4]. We shall make the following assumptions.

(1) F(t, x, u) is a compact set (in R^m) valued function defined in $I \times R^m \times R^r$ where $I = [t_o, t_o + a] \subset R'$.

(2) F(t, x, u) is measurable in t [that is, $\{t : t \in T, F(t,x,u) \subset C\}$ is measurable for every compact set

C in R^m] for each fixed $(x, u) \in R^m \times R^r$, and continuous in (x, u) for each fixed $t \in I$.

(3) $F(t, x, u)$ carries every bounded set in $I \times R^m \times R^r$ into a bounded set in R^m.

(4) $Q(t, x)$ is a compact set (in R^r) valued function defined in $I \times R^m$ and measurable in t for each fixed $x \in R^m$, and upper semicontinuous in x [that is, for every x_o and $\varepsilon > 0$ we can find some neighbourhood V of x_o such that, $N^{\epsilon}(Q(t, x_o)) \quad Q(t,x)$ for every $x \in V$] for each fixed $t \in I$.

(5) $Q(t, x)$ carries every bounded set in $I \times R^m$ into a bounded set in R^r.

(6) $R(t,x) = F(t,x,Q(t,x)) = \{ y : y \in F(t,x,u), u \in Q(t,x)\}$

is a compact and convex set (in R^m) for each $(t,x) \in I \times R^m$.

(7) For every t and x and $u \in Q(t,x)$, $x.y \leq C(|x|^2+1)$ holds for every y such that $y \in F(t,x,u)$, where the dot denotes the scalar product.

(8) K is a compact set in R^m. $K(t)$ is a compact-set (in R^m) valued function defined in I and upper semicontinuous in t.

(9) $f(t,x)$ is a real-valued defined in $I \times R^m$, and is measurable in t for each fixed $x \in R^m$, and continuous in x for each fixed $t \in I$, and is bounded from below.

If $u(t)$ is a measurable function in R^r, $F(t,x,u(t))$ is measurable in t for each fixed $x \in R^m$, and is continuous in x

for each fixed $t \varepsilon I$. Therefore for each measurable function $u(t)$ the system of equations,

$$\dot{x}(t) \varepsilon F(t, x(t), u(t)) \quad \text{for almost all} \quad t \varepsilon I$$

$$x(t_o) = x_o$$

has an absolutely continuous solution $x(t)$ for every $x_o \varepsilon R^m$, if $F(t, x, u)$ satisfies the assumptions stated above.

We say that $x(t)$ is the trajectory corresponding to a control $u(t)$ (measurable in t and $\varepsilon Q(t, x(t))$ on I) if $x(t)$ is an m-dimensional, absolutely continuous function satisfying the above system of equations.

We say that a control $u(t)$, defined for $t_o \leq t \leq \bar{t}$, $\bar{t} \varepsilon I$, transfers K to $K(t)$ if one of the trajectories $x(t)$ corresponding to $u(t)$ satisfies the relations $x(t_o) \varepsilon K$ and $x(\bar{t}) \varepsilon K(\bar{t})$.

We shall consider the problem of finding a control function $u(t)$ which transfers K to $K(t)$ and which minimises the cost functional

$$J(x) = \int_{t_o}^{\bar{t}} f(t, x(t))\, dt$$

where $x(t)$ is one of the solutions corresponding to $u(t)$, and $\bar{t}$ represents a value of t such that $x(t) \varepsilon K(t)$.

<u>Theorem 9.4</u> : Suppose that the conditions stated above are satisfied. Also suppose that there exists at least one control $u(t)$ which transfers K to $K(t)$ on I. Then there exists an

optimal control, that is, a measurable function $u^*(t)$ for which one of the corresponding solutions, $x^*(t)$, with initial condition $x^*(t_o) \varepsilon K$, attains $K(t^*)$ for some $t^* \varepsilon I$, and

$$\inf J(x) = J(x^*) = \int_{t_o}^{t^*} f(t, x^*(t))\, dt,$$

where, in addition, $u^*(t) \varepsilon Q(t, x^*(t))$.

We will indicate the proof of this theorem. Consider the set of all the $x(t)$ satisfying

$$\dot{x}(t) \varepsilon F(t, x(t), u(t)) \text{ almost everywhere on } I,$$

$x(t_o) \varepsilon K$ and $x(\bar{t}) \varepsilon K(\bar{t})$ for some $\bar{t} \varepsilon I$, where, in addition, $u(t) \varepsilon Q(t, x(t))$ for some control $u(t)$. Since one such solution exists by hypothesis, this set is not empty. Consequently we can select a sequence of trajectories $\{x_n(t)\}$ on I, with

$$J(x_n) = \int_{t_o}^{t_n} f(t, x_n(t))\, dt$$

decreasing monotonically to $\inf J(x)$, where t_n represents a value of t such that $x_n(t) \varepsilon K(t)$. $x_n(t)$ satisfy the following relations

$$\dot{x}_n(t) \varepsilon R(t, x_n(t))$$

almost everywhere on I and $x_n(t_o) \varepsilon K$. By the compactness of solutions of the contingent equation [See pp 95-96 in [4] for a proof of the compactness of solutions] we conclude that

$$\dot{x}^*(t) \in R(t, x^*(t)), \quad x^*(t_o) \in K,$$

where $x^*(t)$ is a limit function of a subsequence of $\{x_n(t)\}$. Also we can select a further subsequence (without changing the notation) such that $\{t_n\}$ converges to some $t^* \in I$ since I is a compact interval. Further, making use of the equi-continuity of $\{x_n(t)\}$ and the upper semi-continuity of $K(t)$, we conclude that $x^*(t_o) \in K$ and $x^*(t^*) \in K(t^*)$. Now one can select a measurable function $u^*(t)$ such that

$$\dot{x}^*(t) \in F(t, x^*(t), u^*(t))$$

almost everywhere on I and $u^*(t) \in Q(t, x^*(t))$ on I. Finally

$$J(x_n) = \int_{t_o}^{t_n} f(t, x_n(t))\, dt$$

approaches

$$\int_{t_o}^{t^*} f(t, x^*(t))\, dt$$

as $n \to \infty$ and hence $\inf J(x) = J(x^*)$. Thus $x^*(t)$ on $t_o \leq t \leq t^*$ is optimal.

One could state theorem 9.4 more generally in view of the results in section three but we will not be doing that now. For further results in this connexion refer [5] where the set $R(t, x)$ is not necessarily convex.

(f) Set-valued measures : We shall now discuss about set-valued measures which have many interesting applications in Mathematical economics. See Debreu and Schmeidler [3] and Vind [10].

Let $(T, \underline{\underline{T}})$ be a measurable space, and S be a finite dimensional real vector space. A set-valued measure Φ on $(T, \underline{\underline{T}})$ is a function from T to the non-empty subsets of S, which is countably additive, that is, $\Phi\left(\bigcup_{j=1}^{\infty} E_j\right) = \sum_{j=1}^{\infty} \Phi(E_j)$ for every sequence $E_1, E_2, \ldots,$ of mutually disjoint elements of $\underline{\underline{T}}$. Here the sum $\sum_{j=1}^{\infty} A_j$ of the subsets $A_1, A_2, \ldots,$ of S, consists of all the vectors $a = \sum_{j=1}^{\infty} a_j$, where the series is absolutely convergent, and $a_j \in A_j$ for every $j = 1, 2, \ldots$.

Definition : Let Φ be a set-valued measure on $(T, \underline{\underline{T}})$. A vector-valued measure μ on $(T, \underline{\underline{T}})$ is a selector of Φ if $\mu(E) \in \Phi(E)$ for every $E \in \underline{\underline{T}}$.

We investigate now the existence of selectors, and their properties. In particular, the following problem is interesting : Give conditions implying that for every $E \in \underline{\underline{T}}$ and every $x \in \Phi(E)$ there exists a selector μ of Φ such that $\mu(E) = x$. Recall that Φ is absolutely continuous with respect to the measure λ if $\lambda(E) = 0$ implies $\Phi(E) = \{0\}$. Next we state a theorem due to Artstein [1].

Theorem 9.5 : Let $(T, \underline{\underline{T}}, \lambda)$ be a measure space, λ is finite and non-negative. Let Φ be a set-valued measure, with convex values (that is, for each $E \in \underline{\underline{T}}$, $\Phi(E)$ is a convex subset of S), and such that Φ is absolutely continuous with respect to λ. Then for every $x \in \Phi(E)$ there is a selector μ of Φ such that $\mu(E) = x$.

We will indicate the main steps in the proof. For a detailed proof refer to Artstein's paper.

Note that it is enough to consider the case $E = T$. If $\Phi(T)$ is a singleton then $\Phi(E) = \{\mu(E)\}$ is a singleton for every $E \in \underline{\underline{T}}$, and μ is a measure. Thus μ is the required selector. Proceed by induction on the dimension of $\Phi(T)$, and suppose $\dim \Phi(T) > 0$. Let $x \in \Phi(T)$. If $x \notin$ relative interior of $\Phi(T)$ then for a certain $p \in S$,

$$\begin{aligned} p.x &= \sup \{ p.y : y \in \Phi(T) \} \\ &> \inf \{ p.y : y \in \Phi(T) \} \end{aligned}$$

Now define for every $E \in \underline{\underline{T}}$, $\nu_p(E) = \sup \{ p.x : x \in \Phi(E)\}$ and $\Phi_p(E) = \{ x : x \in \Phi(E)$ and $p.x = \nu_p(E)\}$. Then one can show that Φ_p is a set valued measure. Indeed, since $x \in \Phi_p(T)$ it follows that $\Phi_p(E) \neq \emptyset$ for every E. Obviously, $\Phi_p(T)$ has dimension less than $\Phi(T)$. According to the induction hypothesis there is a selector μ of Φ_p such that $\mu(T) = x$. It is clear that μ is also a selector of Φ.

When $x \in$ relative interior, one can construct a suitable set valued function G such that

$$\int_E G(t)\, d\lambda(t) = \text{relative interior of } \Phi(E) \text{ for every } E \in \underline{\underline{T}}.$$

[Here $\int G(t)\, d\lambda(t)$ stands for the set of all integrable functions f with respect to the measure λ and $f(t) \in G(t)$ for λ-almost all values of t]. In particular, there exists an integrable selection g of G such that $\int_T g(t)\, d\lambda(t) = x$. Define $\mu(E) = \int_E g(t)\, d\lambda(t)$. Then μ is the required selector.

One can give examples to show that theorem 9.5 is false without the convexity assumption or without the condition that ϕ is absolutely continuous with respect to the measure λ.

We will close this section after mentioning the results obtained by L. D. Brown (Personal communication). Let $X = [0,1]$ and Y be a non-metrizable compact space. Let $\underline{\underline{X}}$ and $\underline{\underline{Y}}$ be σ-fields of subsets of X and Y respectively. Of course $\underline{\underline{X}}$ is the usual Borel σ-field containing all closed sets of X. Let $\underline{\underline{C}}$ stand for the product σ-field generated by means of rectangles $A \times B$, $A \in \underline{\underline{X}}$, $B \in \underline{\underline{Y}}$. Let $S \in \underline{\underline{C}}$. If $\underline{\underline{Y}}$ is the Baire σ-field on Y then projection of S onto $[0,1]$ is analytic but the result is false if $\underline{\underline{Y}}$ is the Borel σ-field on Y. If Y is the Baire σ-field on Y and if the continuum hypothesis is assumed to be valid then there exists a measurable selection of S, that is, a measurable function f : Projection $[S] \rightarrow Y$. [I also understand from L.D.Brown's Communication that he is writing a lecture notes on Statistical decision theory which will contain a chapter on applications of selections theorems to decision problems with new proofs of some of the selection theorems].

REFERENCES

[1] Z. Artstein, Set-valued measures, Research Memorandum No. 66, [1971], Department of Mathematics, Hebrew University, Jerusalem, ISRAEL.

[2] H. Bauer and H.S.Bear, The part metric in convex sets, Pac. Jour. Math. 30 [1969], 15-33.

[3] G. Debreu and D. Schmeidler, The Radon-Nikodym derivative of a correspondence, to appear.

[4] N. Kikuchi, Control problems of contingent equation, Publ. RIMS, Kyoto Univ. Ser. A, 3 [1967], 85-99.

[5] N. Kikuchi, On contingent equations, Japan - U.S. seminar on ordinary differential and functional equations, [1971], preprint.

[6] L. F. McAuley and D. F. Addis, Sections and Selections, to appear.

[7] E. Michael, Continuous selections II, Ann. Math 64 [1956], 562-580.

[8] E. Michael, Continuous selections III, Ann. Math 64 [1957], 375-390.

[9] S. N. Patnaik, On the Kakutani fixed point theorem and its relationship with the selection problem I, Springer Verlag Lecture notes in Mathematics, edited by W.M.Fleischman, on 'Set-valued mappings, selections and topological properties of 2^X', No.171, [1970], 70-73.

[10] K. Vind, Edgeworth-allocations in an exchange economy with many traders, _Int. Econ. Rev_ 5 [1964], 165-177.

Lecture Notes in Mathematics

Comprehensive leaflet on request

Vol. 74: A. Fröhlich, Formal Groups. IV, 140 pages. 1968. DM 16,-

Vol. 75: G. Lumer, Algèbres de fonctions et espaces de Hardy. VI, 80 pages. 1968. DM 16,-

Vol. 76: R. G. Swan, Algebraic K-Theory. IV, 262 pages. 1968. DM 18,-

Vol. 77: P.-A. Meyer, Processus de Markov: la frontière de Martin. IV, 123 pages. 1968. DM 16,-

Vol. 78: H. Herrlich, Topologische Reflexionen und Coreflexionen. XVI, 166 Seiten. 1968. DM 16,-

Vol. 79: A. Grothendieck, Categories Cofibrées Additives et Complexe Cotangent Relatif. IV, 167 pages. 1968. DM 16,-

Vol. 80: Seminar on Triples and Categorical Homology Theory. Edited by B. Eckmann. IV, 398 pages. 1969. DM 20,-

Vol. 81: J.-P. Eckmann et M. Guenin, Méthodes Algébriques en Mécanique Statistique. VI, 131 pages. 1969. DM 16,-

Vol. 82: J. Wloka, Grundräume und verallgemeinerte Funktionen. VIII, 131 Seiten. 1969. DM 16,-

Vol. 83: O. Zariski, An Introduction to the Theory of Algebraic Surfaces. Second Printing. IV, 100 pages. 1972. DM 16,-

Vol. 84: H. Lüneburg, Transitive Erweiterungen endlicher Permutationsgruppen. IV, 119 Seiten. 1969. DM 16,-

Vol. 85: P. Cartier et D. Foata, Problèmes combinatoires de commutation et réarrangements. IV, 88 pages. 1969. DM 16,-

Vol. 86: Category Theory, Homology Theory and their Applications I. Edited by P. Hilton. VI, 216 pages. 1969. DM 16,-

Vol. 87: M. Tierney, Categorical Constructions in Stable Homotopy Theory. IV, 65 pages. 1969. DM 16,-

Vol. 88: Séminaire de Probabilités III. IV, 229 pages. 1969. DM 18,-

Vol. 89: Probability and Information Theory. Edited by M. Behara, K. Krickeberg and J. Wolfowitz. IV, 256 pages. 1969. DM 18,-

Vol. 90: N. P. Bhatia and O. Hajek, Local Semi-Dynamical Systems. II, 157 pages. 1969. DM 16,-

Vol. 91: N. N. Janenko, Die Zwischenschrittmethode zur Lösung mehrdimensionaler Probleme der mathematischen Physik. VIII, 194 Seiten. 1969. DM 16,80

Vol. 92: Category Theory, Homology Theory and their Applications II. Edited by P. Hilton. V, 308 pages. 1969. DM 20,-

Vol. 93: K. R. Parthasarathy, Multipliers on Locally Compact Groups. III, 54 pages. 1969. DM 16,-

Vol. 94: M. Machover and J. Hirschfeld, Lectures on Non-Standard Analysis. VI, 79 pages. 1969. DM 16,-

Vol. 95: A. S. Troelstra, Principles of Intuitionism. II, 111 pages. 1969. DM 16,-

Vol. 96: H.-B. Brinkmann und D. Puppe, Abelsche und exakte Kategorien, Korrespondenzen. V, 141 Seiten. 1969. DM 16,-

Vol. 97: S. O. Chase and M. E. Sweedler, Hopf Algebras and Galois Theory. II, 133 pages. 1969. DM 16,-

Vol. 98: M. Heins, Hardy Classes on Riemann Surfaces. III, 106 pages. 1969. DM 16,-

Vol. 99: Category Theory, Homology Theory and their Applications III. Edited by P. Hilton. IV, 489 pages. 1969. DM 24,-

Vol. 100: M. Artin and B. Mazur, Etale Homotopy. II, 196 Seiten. 1969. DM 16,-

Vol. 101: G. P. Szegö et G. Treccani, Semigruppi di Trasformazioni Multivoche. VI, 177 pages. 1969. DM 16,-

Vol. 102: F. Stummel, Rand- und Eigenwertaufgaben in Sobolewschen Räumen. VIII, 386 Seiten. 1969. DM 20,-

Vol. 103: Lectures in Modern Analysis and Applications I. Edited by C. T. Taam. VII, 162 pages. 1969. DM 16,-

Vol. 104: G. H. Pimbley, Jr., Eigenfunction Branches of Nonlinear Operators and their Bifurcations. II, 128 pages. 1969. DM 16,-

Vol. 105: R. Larsen, The Multiplier Problem. VII, 284 pages. 1969. DM 18,-

Vol. 106: Reports of the Midwest Category Seminar III. Edited by S. Mac Lane. III, 247 pages. 1969. DM 16,-

Vol. 107: A. Peyerimhoff, Lectures on Summability. III, 111 pages. 1969. DM 16,-

Vol. 108: Algebraic K-Theory and its Geometric Applications. Edited by R. M. F. Moss and C. B. Thomas. IV, 86 pages. 1969. DM 16,-

Vol. 109: Conference on the Numerical Solution of Differential Equations. Edited by J. Ll. Morris. VI, 275 pages. 1969. DM 18,-

Vol. 110: The Many Facets of Graph Theory. Edited by G. Chartrand and S. F. Kapoor. VIII, 290 pages. 1969. DM 18,-

Vol. 111: K. H. Mayer, Relationen zwischen charakteristischen Zahlen. III, 99 Seiten. 1969. DM 16,-

Vol. 112: Colloquium on Methods of Optimization. Edited by N. N. Moiseev. IV, 293 pages. 1970. DM 18,-

Vol. 113: R. Wille, Kongruenzklassengeometrien. III, 99 Seiten. 1970. DM 16,-

Vol. 114: H. Jacquet and R. P. Langlands, Automorphic Forms on GL (2). VII, 548 pages. 1970. DM 24,-

Vol. 115: K. H. Roggenkamp and V. Huber-Dyson, Lattices over Orders I. XIX, 290 pages. 1970. DM 18,-

Vol. 116: Séminaire Pierre Lelong (Analyse) Année 1969. IV, 195 pages. 1970. DM 16,-

Vol. 117: Y. Meyer, Nombres de Pisot, Nombres de Salem et Analyse Harmonique. 63 pages. 1970. DM 16,-

Vol. 118: Proceedings of the 15th Scandinavian Congress, Oslo 1968. Edited by K. E. Aubert and W. Ljunggren. IV, 162 pages. 1970. DM 16,-

Vol. 119: M. Raynaud, Faisceaux amples sur les schémas en groupes et les espaces homogènes. III, 219 pages. 1970. DM 16,-

Vol. 120: D. Siefkes, Büchi's Monadic Second Order Successor Arithmetic. XII, 130 Seiten. 1970. DM 16,-

Vol. 121: H. S. Bear, Lectures on Gleason Parts. III, 47 pages. 1970. DM 16,-

Vol. 122: H. Zieschang, E. Vogt und H.-D. Coldewey, Flächen und ebene diskontinuierliche Gruppen. VIII, 203 Seiten. 1970. DM 16,-

Vol. 123: A. V. Jategaonkar, Left Principal Ideal Rings. VI, 145 pages. 1970. DM 16,-

Vol. 124: Séminare de Probabilités IV. Edited by P. A. Meyer. IV, 282 pages. 1970. DM 20,-

Vol. 125: Symposium on Automatic Demonstration. V, 310 pages. 1970. DM 20,-

Vol. 126: P. Schapira, Théorie des Hyperfonctions. XI, 157 pages. 1970. DM 16,-

Vol. 127: I. Stewart, Lie Algebras. IV, 97 pages. 1970. DM 16,-

Vol. 128: M. Takesaki, Tomita's Theory of Modular Hilbert Algebras and its Applications. II, 123 pages. 1970. DM 16,-

Vol. 129: K. H. Hofmann, The Duality of Compact Semigroups and C*-Bigebras. XII, 142 pages. 1970. DM 16,-

Vol. 130: F. Lorenz, Quadratische Formen über Körpern. II, 77 Seiten. 1970. DM 16,-

Vol. 131: A. Borel et al., Seminar on Algebraic Groups and Related Finite Groups. VII, 321 pages. 1970. DM 22,-

Vol. 132: Symposium on Optimization. III, 348 pages. 1970. DM 22,-

Vol. 133: F. Topsøe, Topology and Measure. XIV, 79 pages. 1970. DM 16,-

Vol. 134: L. Smith, Lectures on the Eilenberg-Moore Spectral Sequence. VII, 142 pages. 1970. DM 16,-

Vol. 135: W. Stoll, Value Distribution of Holomorphic Maps into Compact Complex Manifolds. II, 267 pages. 1970. DM 18,-

Vol. 136: M. Karoubi et al., Séminaire Heidelberg-Saarbrücken-Strasbourg sur la K-Théorie. IV, 264 pages. 1970. DM 18,-

Vol. 137: Reports of the Midwest Category Seminar IV. Edited by S. MacLane. III, 139 pages. 1970. DM 16,-

Vol. 138: D. Foata et M. Schützenberger, Théorie Géométrique des Polynômes Eulériens. V, 94 pages. 1970. DM 16,-

Vol. 139: A. Badrikian, Séminaire sur les Fonctions Aléatoires Linéaires et les Mesures Cylindriques. VII, 221 pages. 1970. DM 18,-

Vol. 140: Lectures in Modern Analysis and Applications II. Edited by C. T. Taam. VI, 119 pages. 1970. DM 16,-

Vol. 141: G. Jameson, Ordered Linear Spaces. XV, 194 pages. 1970. DM 16,-

Vol. 142: K. W. Roggenkamp, Lattices over Orders II. V, 388 pages. 1970. DM 22,-

Vol. 143: K. W. Gruenberg, Cohomological Topics in Group Theory. XIV, 275 pages. 1970. DM 20,-

Vol. 144: Seminar on Differential Equations and Dynamical Systems, II. Edited by J. A. Yorke. VIII, 268 pages. 1970. DM 20,-

Vol. 145: E. J. Dubuc, Kan Extensions in Enriched Category Theory. XVI, 173 pages. 1970. DM 16,-

Please turn over

Vol. 146: A. B. Altman and S. Kleiman, Introduction to Grothendieck Duality Theory. II, 192 pages. 1970. DM 18,–

Vol. 147: D. E. Dobbs, Cech Cohomological Dimensions for Commutative Rings. VI, 176 pages. 1970. DM 16,–

Vol. 148: R. Azencott, Espaces de Poisson des Groupes Localement Compacts. IX, 141 pages. 1970. DM 16,–

Vol. 149: R. G. Swan and E. G. Evans, K-Theory of Finite Groups and Orders. IV, 237 pages. 1970. DM 20,–

Vol. 150: Heyer, Dualität lokalkompakter Gruppen. XIII, 372 Seiten. 1970. DM 20,–

Vol. 151: M. Demazure et A. Grothendieck, Schémas en Groupes I. (SGA 3). XV, 562 pages. 1970. DM 24,–

Vol. 152: M. Demazure et A. Grothendieck, Schémas en Groupes II. (SGA 3). IX, 654 pages. 1970. DM 24,–

Vol. 153: M. Demazure et A. Grothendieck, Schémas en Groupes III. (SGA 3). VIII, 529 pages. 1970. DM 24,–

Vol. 154: A. Lascoux et M. Berger, Variétés Kähleriennes Compactes. VII, 83 pages. 1970. DM 16,–

Vol. 155: Several Complex Variables I, Maryland 1970. Edited by J. Horváth. IV, 214 pages. 1970. DM 18,–

Vol. 156: R. Hartshorne, Ample Subvarieties of Algebraic Varieties. XIV, 256 pages. 1970. DM 20,–

Vol. 157: T. tom Dieck, K. H. Kamps und D. Puppe, Homotopietheorie. VI, 265 Seiten. 1970. DM 20,–

Vol. 158: T. G. Ostrom, Finite Translation Planes. IV, 112 pages. 1970. DM 16,–

Vol. 159: R. Ansorge und R. Hass, Konvergenz von Differenzenverfahren für lineare und nichtlineare Anfangswertaufgaben. VIII, 145 Seiten. 1970. DM 16,–

Vol. 160: L. Sucheston, Contributions to Ergodic Theory and Probability. VII, 277 pages. 1970. DM 20,–

Vol. 161: J. Stasheff, H-Spaces from a Homotopy Point of View. VI, 95 pages. 1970. DM 16,–

Vol. 162: Harish-Chandra and van Dijk, Harmonic Analysis on Reductive p-adic Groups. IV, 125 pages. 1970. DM 16,–

Vol. 163: P. Deligne, Equations Différentielles à Points Singuliers Reguliers. III, 133 pages. 1970. DM 16,–

Vol. 164: J. P. Ferrier, Seminaire sur les Algebres Complètes. II, 69 pages. 1970. DM 16,–

Vol. 165: J. M. Cohen, Stable Homotopy. V, 194 pages. 1970. DM 16,–

Vol. 166: A. J. Silberger, PGL_2 over the p-adics: its Representations, Spherical Functions, and Fourier Analysis. VII, 202 pages. 1970. DM 18,–

Vol. 167: Lavrentiev, Romanov and Vasiliev, Multidimensional Inverse Problems for Differential Equations. V, 59 pages. 1970. DM 16,–

Vol. 168: F. P. Peterson, The Steenrod Algebra and its Applications: A conference to Celebrate N. E. Steenrod's Sixtieth Birthday. VII, 317 pages. 1970. DM 22,–

Vol. 169: M. Raynaud, Anneaux Locaux Henséliens. V, 129 pages. 1970. DM 16,–

Vol. 170: Lectures in Modern Analysis and Applications III. Edited by C. T. Taam. VI, 213 pages. 1970. DM 18,–

Vol. 171: Set-Valued Mappings, Selections and Topological Properties of 2^X. Edited by W. M. Fleischman. X, 110 pages. 1970. DM 16,–

Vol. 172: Y.-T. Siu and G. Trautmann, Gap-Sheaves and Extension of Coherent Analytic Subsheaves. V, 172 pages. 1971. DM 16,–

Vol. 173: J. N. Mordeson and B. Vinograde, Structure of Arbitrary Purely Inseparable Extension Fields. IV, 138 pages. 1970. DM 16,–

Vol. 174: B. Iversen, Linear Determinants with Applications to the Picard Scheme of a Family of Algebraic Curves. VI, 69 pages. 1970. DM 16,–

Vol. 175: M. Brelot, On Topologies and Boundaries in Potential Theory. VI, 176 pages. 1971. DM 18,–

Vol. 176: H. Popp, Fundamentalgruppen algebraischer Mannigfaltigkeiten. IV, 154 Seiten. 1970. DM 16,–

Vol. 177: J. Lambek, Torsion Theories, Additive Semantics and Rings of Quotients. VI, 94 pages. 1971. DM 16,–

Vol. 178: Th. Bröcker und T. tom Dieck, Kobordismentheorie. XVI, 191 Seiten. 1970. DM 18,–

Vol. 179: Seminaire Bourbaki – vol. 1968/69. Exposés 347-363. IV, 295 pages. 1971. DM 22,–

Vol. 180: Seminaire Bourbaki – vol. 1969/70. Exposés 364-381. IV, 310 pages. 1971. DM 22,–

Vol. 181: F. DeMeyer and E. Ingraham, Separable Algebras over Commutative Rings. V, 157 pages. 1971. DM 16,–

Vol. 182: L. D. Baumert, Cyclic Difference Sets. VI, 166 pages. 1971. DM 16,–

Vol. 183: Analytic Theory of Differential Equations. Edited by P. F. Hsieh and A. W. J. Stoddart. VI, 225 pages. 1971. DM 20,–

Vol. 184: Symposium on Several Complex Variables, Park City, Utah, 1970. Edited by R. M. Brooks. V, 234 pages. 1971. DM 20,–

Vol. 185: Several Complex Variables II, Maryland 1970. Edited by J. Horváth. III, 287 pages. 1971. DM 24,–

Vol. 186: Recent Trends in Graph Theory. Edited by M. Capobianco/J. B. Frechen/M. Krolik. VI, 219 pages. 1971. DM 18,–

Vol. 187: H. S. Shapiro, Topics in Approximation Theory. VIII, 275 pages. 1971. DM 22,–

Vol. 188: Symposium on Semantics of Algorithmic Languages. Edited by E. Engeler. VI, 372 pages. 1971. DM 26,–

Vol. 189: A. Weil, Dirichlet Series and Automorphic Forms. V, 164 pages. 1971. DM 16,–

Vol. 190: Martingales. A Report on a Meeting at Oberwolfach, May 17-23, 1970. Edited by H. Dinges. V, 75 pages. 1971. DM 16,–

Vol. 191: Séminaire de Probabilités V. Edited by P. A. Meyer. IV, 372 pages. 1971. DM 26,–

Vol. 192: Proceedings of Liverpool Singularities – Symposium I. Edited by C. T. C. Wall. V, 319 pages. 1971. DM 24,–

Vol. 193: Symposium on the Theory of Numerical Analysis. Edited by J. Ll. Morris. VI, 152 pages. 1971. DM 16,–

Vol. 194: M. Berger, P. Gauduchon et E. Mazet. Le Spectre d'une Variété Riemannienne. VII, 251 pages. 1971. DM 22,–

Vol. 195: Reports of the Midwest Category Seminar V. Edited by J. W. Gray and S. Mac Lane. III, 255 pages. 1971. DM 22,–

Vol. 196: H-spaces – Neuchâtel (Suisse) – Août 1970. Edited by F. Sigrist. V, 156 pages. 1971. DM 16,–

Vol. 197: Manifolds – Amsterdam 1970. Edited by N. H. Kuiper. V, 231 pages. 1971. DM 20,–

Vol. 198: M. Hervé, Analytic and Plurisubharmonic Functions in Finite and Infinite Dimensional Spaces. VI, 90 pages. 1971. DM 16,–

Vol. 199: Ch. J. Mozzochi, On the Pointwise Convergence of Fourier Series. VII, 87 pages. 1971. DM 16,–

Vol. 200: U. Neri, Singular Integrals. VII, 272 pages. 1971. DM 22,–

Vol. 201: J. H. van Lint, Coding Theory. VII, 136 pages. 1971. DM 16,–

Vol. 202: J. Benedetto, Harmonic Analysis on Totally Disconnected Sets. VIII, 261 pages. 1971. DM 22,–

Vol. 203: D. Knutson, Algebraic Spaces. VI, 261 pages. 1971. DM 22,–

Vol. 204: A. Zygmund, Intégrales Singulières. IV, 53 pages. 1971. DM 16,–

Vol. 205: Séminaire Pierre Lelong (Analyse) Année 1970. VI, 243 pages. 1971. DM 20,–

Vol. 206: Symposium on Differential Equations and Dynamical Systems. Edited by D. Chillingworth. XI, 173 pages. 1971. DM 16,–

Vol. 207: L. Bernstein, The Jacobi-Perron Algorithm – Its Theory and Application. IV, 161 pages. 1971. DM 16,–

Vol. 208: A. Grothendieck and J. P. Murre, The Tame Fundamental Group of a Formal Neighbourhood of a Divisor with Normal Crossings on a Scheme. VIII, 133 pages. 1971. DM 16,–

Vol. 209: Proceedings of Liverpool Singularities – Symposium II. Edited by C. T. C. Wall. V, 280 pages. 1971. DM 22,–

Vol. 210: M. Eichler, Projective Varieties and Modular Forms. III, 118 pages. 1971. DM 16,–

Vol. 211: Théorie des Matroïdes. Edité par C. P. Bruter. III, 108 pages. 1971. DM 16,–

Vol. 212: B. Scarpellini, Proof Theory and Intuitionistic Systems. VII, 291 pages. 1971. DM 24,–

Vol. 213: H. Hogbe-Nlend, Théorie des Bornologies et Applications. V, 168 pages. 1971. DM 18,–

Vol. 214: M. Smorodinsky, Ergodic Theory, Entropy. V, 64 pages. 1971. DM 16,–